Lecture Notes in Economics and Mathematical Systems 537

Springer
Berlin
Heidelberg
New York
Hong Kong
London
Milan
Paris
Tokyo

Angelika Esser

Pricing in (In)Complete Markets

Structural Analysis and Applications

Springer

Author

Angelika Esser
Johann Wolfgang Goethe-University Frankfurt
Faculty of Economics and Business Administration
Chair of Derivatives and Financial Engineering
Mertonstraße 17–21
60054 Frankfurt
Germany

Cataloging-in-Publication Data applied for

Bibliographic information published by Die Deutsche Bibliothek.
Die Deutsche Bibliothek lists this publication in the Deutsche Nationalbibliographie; detailed bibliographic data is available in the Internet at <http://dnb.ddb.de>.

ISSN 0075-8442
ISBN 3-540-20817-8 Springer-Verlag Berlin Heidelberg New York

Springer-Verlag is a part of Springer Science+Business Media

springeronline.com

Typesetting: Camera ready by author
Cover design: *Erich Kirchner*, Heidelberg

Printed on acid-free paper 55/3142/du 5 4 3 2 1 0

To my parents

Preface

This book has been developed during my work as a research assistant at the Chair of Derivatives and Financial Engineering, Goethe-University Frankfurt am Main, Germany. It was accepted as a Ph.D. thesis, titled "Pricing in (In)Complete Markets: Structural Analysis and Applications," at the Faculty of Economics and Business Administration of Goethe-University in May 2003.

It is a pleasure to thank all people who helped me with this project during the last five years. First of all, I would like to mention how much my parents and my friends have supported me. The whole group of people I work with have constantly and generously shared their knowledge with me such that I could get a deeper insight into economic questions.

I am especially indebted to my advisor, Professor Christian Schlag, who not only taught me the fundamentals in derivative pricing, but also guided me in understanding current research and encouraged me to develop my own ideas for further work.

The research atmosphere in Frankfurt is very productive, I have received valuable comments from a large number of people. I wish to thank in particular Michael Belledin who supported me during my first year in the derivatives group, my colleagues Burkart Mönch and Dr. Nicole Branger for their long-term collaboration on several research projects, and my colleagues Christoph Benkert and Micong Klimes for instructive discussions.

Finally, I would like to express my deeply felt gratitude to several people from the Faculty of Mathematics, especially to Professor Anton Wakolbinger, Dr. Matthias Birkner, and Dr. Roderich Tumulka for their profound mathematical advice.

Financial support by DZ BANK-Stiftung for the publication of this monograph is gratefully acknowledged.

Frankfurt am Main, November 2003 *Angelika Esser*

Contents

1

Motivation and Overview

The entirely theoretical work presented focuses on structural concepts of no-arbitrage pricing of contingent claims and on modeling aspects of asset price dynamics that are affected by stochastic liquidity. In order to clarify the theoretical findings, we present several applications and illustrative examples including well-known scenarios such as the Black–Scholes [7] (henceforth BS) setting.

A prominent class of models, developed in the late eighties, are extensions of BS adding stochastic factors such as stochastic interest rates, jumps or stochastic volatility (henceforth SV). Mainly, this research has been inspired by the fact that the classical BS model is not able to reproduce the empirical phenomenon of the volatility smile. The class of SV models opened a wide field for theoretical and empirical research over the last decade (e.g. Heston [28], Schöbel and Zhu [37], Bakshi, Cao, and Chen [2], Belledin and Schlag [5]). The theoretical extensions contain a huge class of models considering mostly the pricing of plain vanilla options, while the pricing of non-standard contracts in SV models has not received too much attention. An exception is the work by Zhu [43]. We examine certain exotic contracts in SV models as one of the applications in this thesis. In particular, a quasi-closed form solution for three types of power options is newly derived in this setup.

Another topic that has recently gained in importance is liquidity. A potential lack of liquidity is an important source of risk for portfolio managers, especially when markets collapse. In these situations institutional investors are typically forced to close their positions immediately due to internal or external regulations. If the positions under consideration are substantial in the sense that the trading activity of the institutional investor influences the asset price process considerably, ignoring liquidity issues will result in a dangerous underestimation of the downside risk of a position. To incorporate liquidity into asset pricing models is challenging and has been investigated in the recent literature (see, e.g. Frey [22], Kampovsky and Trautmann [29], Liu and Yong [31], Schönbucher and Wilmott [38]). In this context the modeling of stock price dynamics, including trading strategies of large investors and liquidity aspects, is of major interest. We extend a deterministic liquidity model, which takes the trading strategy of a large investor into account, by introducing a stochastic process for liquidity.

These applications show that extending BS by quantities that are not directly traded such as stochastic volatility or liquidity is responsible for the incompleteness of the market – given they do not follow the same source of risk as the underlying assets. Thus, incomplete setups are natural setups for financial markets.

They have become an interesting issue to investigate ever since attempts have been made to generalize the standard BS model.

Both applications are based on the fundamental theory of no-arbitrage pricing. We therefore provide the basic concepts of martingale pricing. However, the structural aspects we illustrate in this framework are of interest in their own right.

Firstly, without explicit specification of a model setup, we are able to derive an elegant and compact pricing equation similar to BS for (almost) arbitrary payoff types in terms of artificial probabilities. This formula eliminates the need for numerical simulations even in complex models and is applied to power options.

Secondly, we investigate the comparison between discrete and continuous (i.e. diffusion) models. On the one hand, diffusion models represent a class of widely used processes to model asset price dynamics, being applied to problems in mathematical finance. On the other hand, for certain questions it can be more adequate to choose a discrete setup, since prices move discretely in reality. Diffusion models could then be thought of as an appropriate limit of the real world. The question that arises is to what extent the discrete setup allows for an analog treatment and provides comparable results. Furthermore, the general valuation principles for diffusion models – such as normalized price processes and martingale measures – are applied to the investigation of power options and stochastic liquidity.

The thesis consists of two parts: The first part, Chaps. 2 and 3, deals with general theoretical problems in the framework of martingale pricing. The second part, Chaps. 4 and 5, discusses two applications in the context of incomplete markets, namely stochastic volatility and stochastic liquidity.

Chapter 2 first briefly reviews the basic principles of no-arbitrage pricing, namely the absence of arbitrage and market (in)completeness. In an arbitrage-free market there is no opportunity to make a riskless profit without investing capital. The concept of (in)completeness of the market refers to the question of whether a claim is attainable. This is the case when its payoff structure can be replicated by a self-financing portfolio strategy, i.e. a strategy where the numbers of assets held in the portfolio are dynamically adjusted so that no money is withdrawn or injected intermediately. If and only if every claim is attainable, the market is said to be complete.

The principle of no arbitrage is essentially equivalent to the *existence* of an equivalent martingale measure (henceforth EMM) corresponding to an attainable numeraire, i.e. an attainable asset with a positive price process. The *uniqueness* of the EMM is equivalent to no arbitrage and completeness. However, equivalent *martingale* measures are not the only relevant measures. Besides these, we deal with the more general concept of equivalent probability measures. The change between any two equivalent probability measures is described by the Radon–Nikodym density. In the case of two equivalent *martingale* measures the density is, up to a constant, given by the ratio of the corresponding numeraires. We review the computation of the BS formula to motivate the different choices of numeraires in order to obtain an elegant derivation of the closed-form solution. However, for more complex payoff structures it may be the case that no choice of numeraire yields the desired result, namely representing the pricing equation in terms of EMM. Thus, the aim is to find another tool to derive a compact pricing formula similar to BS for an arbitrary payoff. The key idea is to take an appropriate Radon–Nikodym density to change the measure using the more general concept of equivalent measures instead of EMM. Armed with this, we are able to obtain a quasi-closed form pricing equation for a general payoff structure in terms of artificial probabilities instead of martingale measures.

Furthermore, we investigate the question of when a measure, equivalent to the physical measure, is also a martingale measure. In complete markets there always exists an attainable numeraire associated with the measure under consideration such that every equivalent measure is in fact an EMM. This does not prove to be true in incomplete markets. If and only if the density relative to a risk-neutral measure is attainable, the measure is an EMM, namely with respect to the replicating portfolio for the density. The derivations shown are of maximal generality; in particular no specification of models in terms of asset price dynamics is needed. The results are applied to the discussion of power options in Chap. 4 where we again consider a complete and an incomplete market as examples.

The second topic of the theoretical part of the thesis is the structural comparison of continuous and discrete setups, which are discussed in Chap. 3. In comparing several concepts we allow for complete and incomplete markets. First, we describe the dynamics of the exogenously given processes, which represent basis assets as well as state variables. The structure of the dynamics in the discrete setup can be written similarly to the continuous setup using the martingale representation property. Thereby, we replace the Brownian motion by so-called basis martingales as the sources of risk.

Then, we apply the change of measure discussed in Chap. 2 to the diffusion setup following Girsanov's theorem, and to the discrete model proposing a discrete analogue. In this context significant differences become apparent. For every process describing the change of measure Girsanov's theorem provides a probability measure, whereas in the discrete setup signed measures can occur, i.e. measures that assign a negative probability to some event. Moreover, in the discrete model the new basis martingales have a different covariance structure under the new measure compared with the covariance of the original basis martingales under the physical measure. This stands in contrast to the diffusion setup where the new Brownian motion under the new measure has the same features as the original Brownian motion under the physical measure.

Finally, we consider normalized price processes, which are crucial for martingale pricing. We start with risk-neutral measures, which are obtained by selecting the money market account as the numeraire. In principle, we need not alter the derivation of the drift restriction and the market prices of risk when we change the setup from a continuous to a discrete one. We arrive at the risk-neutral measures analogously in both models. However, things are different in the case of arbitrary risky numeraires such as an underlying asset or a portfolio consisting of basis assets. In the diffusion model the features of the asset price dynamics under every EMM, i.e. for an arbitrary numeraire, are the following. The covariance structure remains the same, only the drift changes in a simple way: It is equal to the sum of the short rate of interest and the instantaneous covariation between the return of the basis asset and the return of the numeraire. Unfortunately, the properties of the dynamics in the discrete setup cannot be summarized in such a simple way. Neither for the drift nor for the covariance does there exist a compact structural result. Thus, we see that the general ideas behind the approaches are similar, but not all features of the models coincide. In fact, diffusion models are quite a special class of models with convenient properties. To illustrate the general theory we conclude by presenting a few examples. We show that the standard binomial model fits into our discrete setup. Furthermore, the SV model is considered as an example for an incomplete market within the diffusion setting.

While the first part of the thesis is devoted to general theory, the second part presents two innovative applications in the framework of incomplete markets: Firstly, we deal with exotic options in a stochastic volatility setup and secondly, we introduce a newly developed stochastic liquidity model.

In Chap. 4 we discuss the general pricing formula for three types of power options. The pricing of these options represents an example of exotic contracts since the payoff structure is not piecewise-linear. We build the results on the general theory of change of measure and numeraire given in Chap. 2. The relationship between change of measure and numeraire is again investigated for these special payoff types. Two explicit model setups are considered as examples for the pricing formula. In the context of the BS model, representing a complete market, a detailed insight into the derivation of the results is of major interest: This refers to the change of measure, the calculation of the corresponding numeraire portfolio, and the relation to the dynamics of the normalized stock price process. The second setting under consideration is an SV model serving as an example for an incomplete market. Within this setup the question of attainability of contingent claims is briefly discussed. The key contribution of this chapter is the derivation of a quasi-closed form pricing equation for power options in an SV setup not discussed in the literature so far. The technique used is Fourier inversion following the lines of Schöbel and Zhu [37].

In Chap. 5 we introduce a stochastic liquidity (henceforth SL) model that generalizes a constant liquidity (henceforth CL) model developed by Frey [22]. Adding a stochastic process for liquidity with a second source of risk, allowing for arbitrary correlation between the increments of the Brownian motions, yields an incomplete market. We focus on a scenario where the transactions of a large trader have an impact on prices due to potential illiquidity in the market. This approach regards liquidity as an exogenously given quantity interpreted as the sensitivity of asset prices to transactions of the large trader. Under certain assumptions we are able to derive the effective asset price dynamics in our newly developed model. The dynamics turn out to fit into the class of diffusion models with stochastic volatility. Liquidity has an impact on both the asset price process and the trading strategy. This feature is referred to as the liquidity feedback effect, which we discuss theoretically on the basis of the effective dynamics. One key result is that for generic correlation parameters the volatility increases for every trading strategy used, compared with the CL model, due to the liquidity feedback effect. We consider two types of trading strategies of the large investor, the positive feedback strategy, which means selling (buying) stocks when the asset price drops (rises), and the contrarian feedback strategy, which means selling (buying) when the asset price rises (drops). The implications for the total volatility and the correlation structure are discussed in detail. For sensible parameter choices volatility increases, compared with BS for a positive feedback strategy, similar to the scenario in the CL setup. However, for contrarian strategies the volatility does in general not decrease, compared with BS, as it does in the case of the CL model. This is due to the fact that the liquidity feedback effect can cause destabilizations of the asset price dynamics. Furthermore, we derive the drift restriction and the risk-neutral dynamics in this setup, applying the general theory developed in Chap. 3.

The implications and contributions of the extended liquidity model are illustrated for two examples. First, we introduce two economically motivated trading strategies for the large trader depending on both the underlying asset and liquidity. We use them for Monte Carlo simulations of the asset price processes in the SL setup compared with BS. The simulation studies show that prices in the SL setting may heavily react to changes in liquidity, a feature that cannot be modeled within the CL model. Hence, the new model yields asset price dynamics that take the reaction of prices to liquidity changes explicitly into account. In the second application of the SL model we consider an insurance against the consequences of illiquidity when large positions have to be unwound. The investor may want to be compensated for a price discount in this case. We derive a formula for the arbitrage-free price of such a contract and present a few examples.

Chapter 6 summarizes the key results of the thesis and gives an outlook on future research.

2

Pricing by Change of Measure and Numeraire

2.1 Introduction

There are several ways to derive the no-arbitrage price of a contingent claim, such as following a replicating portfolio strategy or solving a partial differential equation. Another prominent approach is martingale pricing, which is the method we deal with in this chapter. We briefly review well-known facts on equivalent measures, the Radon–Nikodym derivative, martingale measures, and the change of numeraires following Geman, El Karoui, and Rochet [27]. The only measures we consider within this thesis are the ones equivalent to the physical measure. The ultimate goal in deriving the pricing formula for a claim is to write it in terms of possibly different artificial probabilites. It is known that the choice of different numeraires allows for a convenient computation of the claim's fair price. This can be seen when looking at the BS formula: The easiest method for the valuation of a standard call is to choose appropriate normalizing assets and corresponding martingale measures. This will be reviewed to motivate the concept of different numeraires.

However, when the payoff function is more complex this approach no longer works. This is due to the fact that an arbitrary payoff does not usually consist of summands that are attainable assets, which in particular means that they cannot be used as numeraires. Thus, we are looking for a similar, but more general concept using probability measures rather than martingale measures. The first contribution of this chapter is the derivation of a concise pricing equation for a general payoff structure, similar to the pricing equation for standard contracts, which is of particular interest for numerical computations.

Furthermore, the question arises concerning under which circumstances a probability measure, equivalent to the physical measure, is a martingale measure. We will see that in complete markets every measure is a martingale measure, whereas in incomplete markets this does not hold.

2.2 Model Setup

Within this thesis we consider a market with a finite number of basis assets (sometimes referred to as underlying assets), which means that their price processes are exogenously specified. The assets are assumed to have no intermediate payoffs such as dividend payments. There exists at least one risky asset, denoted by S, and a risk-free asset, the money market account (henceforth MMA), denoted by B, with an initial price of one, earning a deterministic short rate of interest. Additionally, there may be a finite number of non-traded state variables, such as stochastic volatility, whose process dynamics are also exogenously given. Furthermore, we consider contingent claims written on the basis assets, denoted by F. Within this thesis we restrict ourselves to European claims with no intermediate payments or exercise properties.

To fix the notation for the rest of the thesis, let $(\Omega, \mathcal{F}, P)$ be a probability space, where Ω denotes the sample space. The filtration $\mathcal{F} = \bigcup_{0 \le t \le T} \mathcal{F}_t$, where $T < \infty$ is fixed, is assumed to be generated by the exogenously given sources of risk, e.g. by a Brownian motion driving the dynamics of the underlying assets and the state variables. Moreover, we assume that the basis assets and state variables are adapted to the filtration. We can keep them on a very general level in this chapter; they are defined explicitly when needed in Chaps. 3, 4, and 5. The measure P is referred to as the physical measure representing the real-world probability.

As already mentioned, the two important concepts concerning the pricing of contingent claims are the principle of no-arbitrage and market (in)completeness. No-arbitrage is essential for the valuation of claims, and we assume throughout this thesis that the prices of the basis assets are such that the market is arbitrage-free. This means that assets with the same payoff structure must have identical prices. The opportunity to earn money in a riskless way with a zero net investment, which we – loosely – define as an arbitrage opportunity, is excluded. Moreover, negative values for non-negative payoffs are ruled out.

The second relevant concept is the completeness of a market. In a complete market it is possible to replicate *every* claim with an arbitrary payoff using a self-financing portfolio, consisting of a dynamically adjusted linear combination of the basis assets. Recall that self-financing means that at no intermediate point in time is money withdrawn from or injected into the portfolio. In an incomplete market not all claims can be replicated. If and only if a payoff of a claim can be replicated by a self-financing portfolio strategy, is the claim called attainable. Thus, the no-arbitrage price in a complete market can be calculated with the use of a self-financing replicating portfolio only, because every claim is attainable. However, this is generally not an appropriate method of pricing. It is either costly in a multi-period discrete setup or it may just be impossible in a continuous model without further knowledge about the price of the claim. A more elegant method of valuation of a claim is martingale pricing, which is the focus of this chapter.

2.3 Equivalent Measures

We have already pointed out the importance of the change of measure technique for martingale pricing. Now, we will discuss it in detail including the use of a Radon–Nikodym density and the change of numeraires. The latter is a special change between equivalent measures, namely between EMM. Both are relevant techniques to derive a compact pricing equation of a contingent claim using artificial probabilities. Since for martingale pricing the physical measure does not play a role, we consider (artificial) measures equivalent to P, i.e. measures having the same null sets. We start with the Radon–Nikodym derivative describing the change between any two equivalent measures.

2.3.1 Radon–Nikodym Derivative

There is an almost trivial way to obtain an equivalent measure $\bar{P}$ from the physical measure P: We can get $\bar{P}$ by simply multiplying the measure P by a density, i.e. by a positive random variable Y with $\mathbf{E}_0^P Y = 1$. It can easily be seen that this yields an equivalent probability measure $\bar{P}$. The converse holds as well according to the theorem of Radon–Nikodym (cf., inter alia, Breiman [10]). Let $\bar{P}$ be an arbitrary probability measure equivalent to P.

Then, the change of measure can be described by a positive random variable Y, the Radon–Nikodym derivative, i.e. $\exists Y > 0$ such that

$$d\bar{P} = Y\, dP \quad \text{on } \mathcal{F}_T. \tag{2.1}$$

We set $\xi_T := Y$ and define the process ξ by $\xi_t := \mathbf{E}_t^P[\xi_T]$, where the subscript t denotes the expectation, conditional on $\mathcal{F}_t$. According to the tower law, ξ is a positive martingale with $\xi_0 = 1$. Conditional on $\mathcal{F}_t$, the relation between the expectations under the two equivalent measures of any terminal payoff $\bar{Y}$ adapted to $\mathcal{F}_T$ is given by

$$\mathbf{E}_t^{\bar{P}}(\bar{Y}) = \mathbf{E}_t^P\left[\frac{\xi_T}{\xi_t}\bar{Y}\right] \tag{2.2}$$

(see, e.g. Baxter and Rennie [4]). The change-of-measure technique will turn out to be extremely useful in the following, and it will be applied to the continuous and discrete setup in Chap. 3.

2.3.2 Martingale Measures

We now focus on a special class of equivalent measures, namely martingale measures with respect to a numeraire, which are the essential ingredients for martingale pricing. To do so, we first present two definitions:

1. An attainable asset N with a positive price process is called a *numeraire.*
2. A measure P', equivalent to the physical measure P, is called an *equivalent martingale measure with respect to the numeraire* N if the price processes of all basis assets, normalized by the numeraire N, are P'-martingales, i.e. if the following martingale pricing formula is true for every basis asset S:

$$\frac{S_t}{N_t} = \mathbf{E}_t^{P'}\left[\frac{S_T}{N_T}\right]. \tag{2.3}$$

While the definitions of *numeraires* differ subtly in the literature (cf., inter alia, Björk [6] and Duffie [18]), the definition above is the most adequate for our questions under consideration.

We further use the two central theorems of no-arbitrage pricing (see, e.g. Duffie [18]):

1. For every numeraire N an equivalent martingale measure exists, if and only if there are no-arbitrage opportunities in the market.[1]

[1] For an appropriate definition of no-arbitrage excluding pathological scenarios see Delbaen and Schachermayer [15].

2. The equivalent martingale measure corresponding to N is unique, if and only if the market is arbitrage-free and complete.

It is well known from general theory (cf., inter alia, Musiela and Rutkowski [33]) that the martingale property in (2.3) does not only hold for the normalized basis assets, but for any other normalized attainable derivative contract as well. This follows from the linearity of the self-financing replicating portfolio (restricting the class of admissible trading strategies appropriately) and the linearity of conditional expectations for attainable claims. Thus, every attainable claim has a unique arbitrage-free price which can be computed from the replicating portfolio without using the martingale pricing formula. This is different for non-attainable claims, i.e. in an incomplete market.

The first theorem implies the following: If there is an arbitrage-free price system for all claims F, then there exists an EMM P' with numeraire N such that the martingale pricing formula holds for every F:

$$\frac{F_t}{N_t} = \mathbf{E}_t^{P'} \left[\frac{F_T}{N_T} \right] . \tag{2.4}$$

Since in an incomplete market the EMM corresponding to a numeraire is not unique, the price of every non-attainable claim depends on the chosen measure. Fixing a measure means fixing a price system free of arbitrage opportunities.

For claims which cannot be replicated the no-arbitrage price lies between two bounds (see, e.g. Musiela and Rutkowski [33]). In certain setups it can be shown that these bounds are sharp, in the sense that a claim can take on all prices within the model-independent lower and upper bounds (see, e.g. Eberlein and Jacod [19], Frey and Sin [24], Branger, Esser, and Schlag [9]).

The selection of the EMM in an incomplete market, taking into account additional assumptions to fix the degrees of freedom, is in general a difficult problem. There are many criteria on which the choice can be based, for instance on investor's preferences in an equilibrium context (see Cox, Ingersoll, and Ross [11]). The issue of which criterion seems most appropriate for the given problem is not discussed here, as it certainly represents a research topic in its own right.

In the set of EMM the risk-neutral measures play a special role. These are the measures under which the discounted price processes are martingales, i.e. they refer to the MMA, denoted by B, as the numeraire. In a complete market the risk-neutral measure is unique, and we denote it by $\hat{P}$. In an incomplete market there exists a class of risk-neutral measures, from which we fix the one corresponding to the given arbitrage-free price system, and call it $\hat{P}$ as well. The pricing formula (2.4) becomes simpler using the risk-neutral measure, since the MMA is assumed to be a deterministic asset:

$$\frac{F_t}{B_t} = \frac{1}{B_T}\hat{\mathbf{E}}_t\left[F_T\right]. \tag{2.5}$$

2.3.3 Change of Numeraire

Since the pricing equation admits a representation for every numeraire and corresponding EMM, we now ask for the relationship between different numeraires. Under the assumption of no-arbitrage the choice of the EMM and the corresponding numeraire must not have an impact on the price of the claim F. As an example, we consider the pairs (P', N) and $(\hat{P}, B)$. Above, the pricing formula for F has been stated twice, first using the pair (P', N) in (2.4) and second using $(\hat{P}, B)$ in (2.5). From this we can derive the relation between P' and $\hat{P}$ conditional on $\mathcal{F}_t$. Since

$$N_t\mathbf{E}_t^{P'}\left[\frac{F_T}{N_T}\right] = \hat{\mathbf{E}}_t\left[\frac{N_tF_T}{N_T}\frac{N_T/N_t}{B_T/B_t}\right] = \frac{B_t}{B_T}\hat{\mathbf{E}}_t\left[F_T\right],$$

it holds that

$$\mathbf{E}_t^{P'}\left[\bar{Y}\right] = \hat{\mathbf{E}}_t\left[\bar{Y}\,\frac{N_T/N_t}{B_T/B_t}\right] \quad \forall\,\bar{Y} \text{ adapted to } \mathcal{F}_T.$$

The fact that the change between EMM is given in terms of fractions of corresponding numeraires is a result by Geman, El Karoui, and Rochet [27]. Moreover, they have proved the following: Let M be an asset with a positive price process such that M/N is a P'-martingale. Then, every asset normalized with M is a martingale under the measure P'', which is defined by

$$dP'' = \frac{M_T/M_0}{N_T/N_0}\,dP' \text{ on } \mathcal{F}_T. \tag{2.6}$$

As an immediate consequence of formula (2.6) we see that the fraction $\frac{M_T/M_0}{N_T/N_0}$ is a positive random variable with an expectation of one. This is due to the fact that M/N is a P'-martingale which implies

$$\frac{M_T/M_0}{N_T/N_0} = \frac{M_T/N_T}{\mathbf{E}_0^{P'}\left[M_T/N_T\right]}. \tag{2.7}$$

Hence, the ratio of numeraires serves indeed as a density so that P'' is a probability measure equivalent to P'. To derive the formula analogously to (2.6) conditional on $\mathcal{F}_t$, we use (2.2) with

$$\xi_t = \mathbf{E}_t^{P'}\left[\frac{M_T/M_0}{N_T/N_0}\right] = \frac{M_t/M_0}{N_t/N_0},$$

which gives

$$\mathbf{E}_t^{P''}\left[\bar{Y}\right] = \mathbf{E}_t^{P'}\left[\bar{Y}\,\frac{M_T/M_t}{N_T/N_t}\right] \qquad \forall \bar{Y} \text{ adapted to } \mathcal{F}_T. \tag{2.8}$$

The martingale pricing formula, being valid for complete and incomplete markets, can then be stated for every claim F using different numeraires N and M as

$$\begin{aligned} F_t &= N_t \mathbf{E}_t^{P'}\left[\frac{F_T}{N_T}\right] \\ &= M_t \mathbf{E}_t^{P''}\left[\frac{F_T}{M_T}\right]. \end{aligned}$$

We now want to clarify the advantage of different choices of numeraires. Why bother with different numeraires at all?

In the sequel it will become obvious that a clever choice of the numeraire asset facilitates the derivation of pricing equations. We consider the derivation of the BS pricing equation, following for instance Musiela and Rutkowski [33].

The terminal payoff of a European call C on the underlying asset S with strike K and maturity T is given by

$$C_T := \max\left[S_T - K; 0\right] = (S_T - K)I(S_T > K),$$

where $I(A)$ denotes the indicator variable of the event A. Using the risk-neutral measure, i.e. the MMA as the numeraire, yields the pricing formula

$$C_t = \frac{B_t}{B_T}\hat{\mathbf{E}}_t\left[S_T I(S_T > K)\right] - \frac{B_t}{B_T}\hat{\mathbf{E}}_t\left[K I(S_T > K)\right]. \tag{2.9}$$

The valuation of the second summand reduces to computing the $\hat{P}$-probability of the corresponding event. For the first summand we take the underlying asset as the numeraire employing a change of measure by the ratio of the numeraires as in Geman, El Karoui, and Rochet [27], similar to (2.8):

$$\tilde{\mathbf{E}}_t\left[\bar{Y}\right] = \hat{\mathbf{E}}_t\left[\bar{Y}\,\frac{S_T/S_t}{B_T/B_t}\right] \qquad \forall \bar{Y} \text{ adapted to } \mathcal{F}_T, \tag{2.10}$$

where $\tilde{P}$ denotes the EMM corresponding to the underlying asset as the numeraire. Then, the first summand simplifies to

$$\begin{aligned}\frac{B_t}{B_T}\hat{\mathbf{E}}_t\left[S_T I(S_T > K)\right] &= B_t\tilde{\mathbf{E}}_t\left[\frac{(B_T/B_t)}{(S_T/S_t)}\frac{S_T}{B_T}I(S_T > K)\right] \\ &= S_t\tilde{P}_t(S_T > K),\end{aligned} \tag{2.11}$$

and the pricing equation becomes

$$C_t = S_t\tilde{P}_t(S_T > K) - \frac{B_t}{B_T}K\hat{P}_t(S_T > K). \tag{2.12}$$

Equation (2.12) is more intuitive and handy compared with (2.9). What is left to calculate for the first term is the artificial probability $\tilde{P}$ – corresponding to the underlying asset as the numeraire – of the event that the option is exercised, instead of the expectation $\hat{\mathbf{E}}_t\left[S_T I(S_T > K)\right]$. The computation of the expectation is more tedious compared with the calculation of the artificial probability. Furthermore, (2.12) already represents a quasi-closed form solution. This simplification has been obtained by switching from the numeraire B to the numeraire S, which shows that a clever choice of numeraire does indeed allow for an elegant derivation of the pricing equation.

But does this approach always work? What happens if S_T in the terminal payoff is substituted by $g(S_T)$, where g is an arbitrary positive function, or by an arbitrary adapted payoff $\mathcal{W}$? Then, the technique of changing numeraires might no longer be applicable, since an arbitrary payoff is usually not attainable and therefore cannot serve as a numeraire. However, we can still change the measure in a clever way using an appropriate Radon–Nikodym derivative to obtain an elegant representation of the pricing equation. This is discussed in detail in the next section.

2.4 Derivation of a General Pricing Equation

The aim of this section is to develop a pricing equation similar to the BS equation (2.12) in terms of artificial probabilities times today's prices. First of all, such a formula is more concise than the pricing formula using expectations. Moreover, a quasi-closed form solution in terms of artificial probabilities is by far preferable, compared with the computation of expectations. This is due to the fact that the latter requires costly numerical simulation techniques in complex scenarios.

We consider an arbitrage-free market, be it complete or not. We look at a claim F with a terminal payoff at T of the form

$$F_T = \mathcal{W} I(f(S_T) > 0),$$

where $\mathcal{W} > 0$, adapted to $\mathcal{F}_T$, and f is arbitrary. This is a typical payoff structure of non-standard European derivatives written on the underlying asset S.

The general pricing formula is given by (2.4) yielding

$$F_t = N_t \mathbf{E}_t^{P'} \left[\frac{\mathcal{W} I(f(S_T) > 0)}{N_T} \right].$$

Now, the aim is to change the measure such that we can rewrite the pricing equation as the product of the artificial probability of the event $f(S_T) > 0$ and today's price of the payoff $\mathcal{W}$. In general, the payoff itself cannot be used as a numeraire, since numeraires are defined as attainable assets. Thus, we are not able to apply the change of numeraire as shown for the plain vanilla call option. However, it is still possible to change the measure choosing an appropriate Radon–Nikodym derivative, in order to achieve the desired representation of the pricing equation. Analogously to the right-hand side of (2.7) we define the measure $\bar{P}$ by the density

$$\frac{d\bar{P}}{dP'} = \frac{\mathcal{W}/N_T}{\mathbf{E}_0^{P'}\left[\mathcal{W}/N_T\right]} \text{ on } \mathcal{F}_T. \tag{2.13}$$

Note that replacing $\mathcal{W}$ by an attainable asset S_T, which can be used as a numeraire, the right-hand side of formula (2.13) simplifies to the ratio of corresponding numeraires, i.e.

$$\frac{S_T/N_T}{\mathbf{E}_0^{P'}\left[S_T/N_T\right]} = \frac{S_T/S_0}{N_T/N_0},$$

since S/N is a P'-martingale.

The pricing equation using the density in (2.13) is then given by

$$\begin{aligned} F_t &= N_t \mathbf{E}_t^{P'} \left[\frac{F_T}{N_T} \right] \\ &= N_t \mathbf{E}_t^{P'} \left[\mathcal{W}/N_T\right] \mathbf{E}_t^{P'} \left[\frac{\mathcal{W}/N_T}{\mathbf{E}_t^{P'}\left[\mathcal{W}/N_T\right]} I(f(S_T) > 0) \right] \\ &= N_t \mathbf{E}_t^{P'} \left[\mathcal{W}/N_T\right] \mathbf{E}_t^{\bar{P}} \left[I(f(S_T) > 0) \right]. \end{aligned}$$

Thus, we get the following representation of the pricing formula for non-negative payoffs F_T of the form $F_T = \mathcal{W}I(f(S_T) > 0)$:

$$F_t = N_t \mathbf{E}_t^{P'} [\mathcal{W}/N_T] \bar{P}_t(f(S_T) > 0). \tag{2.14}$$

The price of the claim F is equal to the product of the artificial probability $\bar{P}$ of the event $f(S_T > 0)$ and today's price of $\mathcal{W}$, given by $N_t \mathbf{E}_t^{P'} [\mathcal{W}/N_T]$. Formula (2.14) holds for an arbitrary, positive $\mathcal{W}$, adapted to $\mathcal{F}_T$, and can therefore be applied to a great class of payoffs. In an incomplete market the price of $\mathcal{W}$ may depend on the choice of the measure P' with respect to N. This is the case whenever the payoff $\mathcal{W}$ is not attainable.

Changing the measure with respect to the risk-neutral measure, we are able to further simplify the density. Normalization with the numeraire (which is the MMA in this case) cancels out, since we have assumed the MMA to be deterministic. The change of measure is then simply given by

$$\frac{d\bar{P}}{d\hat{P}} = \frac{\mathcal{W}}{\hat{\mathbf{E}}_0[\mathcal{W}]}.$$

What payoff types denoted by $\mathcal{W}$ should we consider? For instance, one could think of $\mathcal{W}$ as a positive function of the terminal payoff, i.e. $\mathcal{W} = g(S_T)$. In this case the change of measure is given by

$$\frac{d\bar{P}}{dP'} = \frac{g(S_T)/N_T}{\mathbf{E}_0^{P'} [g(S_T)/N_T]} \quad \text{on } \mathcal{F}_T, \tag{2.15}$$

and the representation of the pricing equation of the claim F with payoff $F_T = g(S_T)I(f(S_T) > 0)$ by

$$F_t = N_t \mathbf{E}_t^{P'} [g(S_T)/N_T] \bar{P}_t(f(S_T) > 0). \tag{2.16}$$

It is important to realize that $g(S_t) \neq N_t \mathbf{E}_t^{P'} [g(S_T)/N_T]$ for generic $g > 0$. This means that the function of the stock price process today is in general not equal to the no-arbitrage price of the payoff $g(S_T)$, because $(g(S_t)/N_t)$ and (S_t/N_t) cannot be martingales simultaneously for non-linear functions g. This will be discussed in detail in Chap. 4 for $g(S) = S^a$. When we consider $g(S) = S$ as in the BS formula, the pricing equation simplifies to

$$F_t = S_t \bar{P}_t(f(S_T) > 0). \tag{2.17}$$

2.5 Is Every Equivalent Measure a Martingale Measure?

This section discusses whether every measure, equivalent to the physical measure, is an EMM with respect to an attainable numeraire. To examine this, we distinguish between complete and incomplete markets. If the equivalent measure is an EMM, the pricing equation (2.14) can be simplified further with the help of the numeraire which is discussed in 2.5.3.

2.5.1 Complete Market

We will see in the following that in a complete setup any change of measure between equivalent measures, which is in general given by the Radon–Nikodym density, can be written in terms of numeraires. This is due to the fact that any payoff is attainable.

Let $\bar{P}$ be an arbitrary measure equivalent to $\hat{P}$, and define the positive random variable $Y = \frac{d\bar{P}}{d\hat{P}}$. Since any payoff can be replicated by a self-financing portfolio, we can choose a portfolio V such that

$$V_T := Y. \tag{2.18}$$

Since $\hat{\mathbf{E}}_0 V_T = 1$ and $V_0 = \hat{\mathbf{E}}_0[V_T/B_T]$, this yields

$$Y = \frac{V_T}{B_T \hat{\mathbf{E}}_0[V_T/B_T]} = \frac{V_T/V_0}{B_T/B_0} ,$$

recalling that $B_0 = 1$. Hence, the change of measure is given by

$$d\bar{P} = \frac{V_T/V_0}{B_T/B_0} d\hat{P} \text{ on } \mathcal{F}_T,$$

which corresponds exactly to formula (2.6). This shows that $\bar{P}$ is indeed the EMM with respect to the numeraire V. Note that even the physical measure is an EMM, since it is a measure equivalent to $\hat{P}$. Thus, there must exist a numeraire for the physical measure P which is a portfolio proportional to the Radon–Nikodym derivative $dP/d\hat{P}$. The numeraire portfolio corresponding to the physical measure is discussed in an equilibrium context by Long [32].

What have we learned about change of measure and numeraire so far?

It is obvious that the change of numeraires is a change between EMM. However, the converse is true as well in the current setting: We have proved that a change between measures, equivalent to the physical measure, yields an EMM. This is summarized in the following:

Proposition 2.1. *Let $\hat{P}$ be the risk-neutral measure in a complete market. Then, the measure $\bar{P}$, equivalent to $\hat{P}$, where $d\bar{P} = Yd\hat{P}$, is an EMM with respect to the numeraire V, where $V_T \equiv Y$.*

2.5.2 Incomplete Market

In an incomplete market, the EMM is not unique, and not every claim can be replicated. Let $\hat{P}$ be a fixed risk-neutral measure. We take an arbitrary equivalent measure $\bar{P}$, where

$$Y = \frac{d\bar{P}}{d\hat{P}} . \tag{2.19}$$

Now, the question arises whether there is an attainable numeraire N corresponding to $\bar{P}$ such that the price system generated by $(\bar{P}, N)$ is the same as the one generated by $(\hat{P}, B)$. First of all, suppose that there exists a self-financing replicating portfolio V for Y. Then, the same arguments as in 2.5.1 imply that $\bar{P}$ is an EMM with respect to the numeraire V. Thus, if the density is attainable, then $\bar{P}$ is an EMM.

The other implication holds as well: If $\bar{P}$ is an EMM corresponding to the attainable numeraire V generating the same price system as $(\hat{P}, B)$, then one must be able to compute $\bar{P}$ from $\hat{P}$ according to Geman, El Karoui, and Rochet [27] using the ratio of numeraires, i.e.

$$\frac{d\bar{P}}{d\hat{P}} = \frac{V_T/V_0}{B_T/B_0} = Y.$$

Since the MMA is deterministic, it holds that

$$Y = \text{const} \cdot V_T.$$

Thus, Y is proportional to the attainable numeraire V, implying that Y itself must be attainable. The analysis for an incomplete market is summarized in the following:

Proposition 2.2. *Consider an incomplete market, and let $\hat{P}$ denote a fixed EMM corresponding to the MMA as the numeraire. Let $\bar{P}$ be an arbitrary measure equivalent to $\hat{P}$ where $Y = d\bar{P}/d\hat{P}$. If and only if Y is attainable, $\bar{P}$ is an EMM generating the same price system as $(\hat{P}, B)$. In that case, $\bar{P}$ is an EMM with respect to the traded numeraire V, where $V_T = \text{const.} \cdot Y$.*

2.5.3 Review of the Pricing Equation

Let us now go back to the general pricing equation (2.14) to find out when there exists a representation in terms of numeraires. Whether a numeraire corresponding to $\bar{P}$ exists depends on the setup or the payoff, respectively. As seen above, we have to distinguish between complete and incomplete markets or, to put it differently, between attainable and non-attainable payoffs. In a complete market or for an attainable payoff $\mathcal{W}$ in the incomplete setup, formula (2.14) can be simplified using the result derived in 2.5.1 and 2.5.2. Since we can choose a replicating portfolio $V_T \equiv \mathcal{W}$, the martingale pricing formula can be rewritten as

$$\begin{aligned} F_t &= N_t \mathbf{E}_t^{P'} \left[\frac{F_T}{N_T} \right] \\ &= V_t \mathbf{E}_t^{\bar{P}} \left[\frac{\mathcal{W} I(f(S_T) > 0)}{V_T} \right] \\ &= V_t \bar{P}_t (f(S_T) > 0), \end{aligned} \tag{2.20}$$

where $\bar{P}$, as defined in (2.13) and used in (2.14), is an EMM with respect to V. Since $\mathcal{W}$ is an attainable payoff replicated by a portfolio V_T, the price of $\mathcal{W}$ at t is simply given by V_t in this case. Thus, the price of the contingent claim is equal to the product of today's price of the replicating portfolio V_t and the artificial probability $\bar{P}$ – being an EMM with respect to V – of the event $f(S_T) > 0$.

If and only if there is no replicating portfolio for $\mathcal{W}$, then $\bar{P}$ is not an EMM and the pricing equation (2.14) cannot be simplified any more.

2.6 Conclusion

This chapter has summarized the key features of the martingale pricing technique. We have derived a compact form of the pricing equation for general payoffs, that are contingent on a specific event, using equivalent measures. Subsequently, the question has been discussed whether every measure, equivalent to the physical measure, is an EMM with respect to an attainable numeraire. The following two results have been developed: Firstly, in a complete market, every measure, equivalent to the physical measure, is an EMM. Secondly, in an incomplete market, there does not exist an attainable numeraire for every measure, equivalent to the physical measure. However, if and only if the Radon–Nikodym density with respect to a fixed risk-neutral measure is attainable, the equivalent measure is an EMM, namely with respect to a portfolio with terminal price proportional to the density. These results are applied to the valuation of power options in Chap. 4.

3

Comparison of Discrete and Continuous Models

3.1 Introduction

A natural classification of models is to distinguish between discrete and continuous setups. The most prominent representatives are the Cox, Ross, and Rubinstein [13] binomial tree model and the classical BS model, respectively. In this chapter we contrast a discrete against a continuous setting for both complete and incomplete markets. Various examples, including SV models as important representatives of incomplete markets, shed light on the general theory. The discrete models we consider are state- and time-discrete; the continuous models we refer to are diffusion models. The latter are widely studied (see, e.g. Protter [36]), but for certain problems it might be more appropriate to look at a discrete setting, being a better approximation of reality. Furthermore, for complex problems it may be more adequate to start in discrete time, avoiding tedious technical conditions necessary in continuous time. In this context, the question arises whether all properties of the diffusion setup carry over to the discrete scenario. We will see that this is not always the case.

On the one hand, it is shown that the key results and the economic intuition are similar in the continuous and discrete setting: We can formulate the dynamics of the stochastic processes in the discrete model in the same fashion as in the continuous model. Furthermore, the drift restriction involving the market prices of risk that determine the risk-neutral measures will be derived analogously in both models.

On the other hand, there are topics where it is advisable to be careful. When considering the change of measure and normalized price processes, significant differences between the continuous and the discrete setting appear. In diffusion models only the drift of the stochastic processes changes when the measure is changed, whereas in the discrete model the covariance structure changes additionally. In the former, there even exists a simple formula for the drift of every asset under an arbitrary EMM: The drift is equal to the sum of the short rate of interest and the instantaneous covariation between the return of the asset and the return of the numeraire. Unfortunately, such a concise form is not available in the discrete model.

For the discrete setup we follow the lines of Dothan [16]. The classical tree models such as the binomial and trinomial models can be embedded into this scenario. This is exemplified by considering a simple one-period binomial tree. When comparing the discrete to the continuous setting, we can deduce that in fact diffusion models are a special class of models with several properties not shared by other classes of models.

3.2 Dynamics of the Underlying Processes

We start with the definition of the dynamics of the basis assets and state variables in the continuous and the discrete setup. Let $(\Omega, \mathcal{F}, P)$ be a probability space as defined in Chap. 2. We assume that there are n exogenously given stochastic processes $S_t^{(i)}, i = 1, \ldots, n, 0 \leq t \leq T$. For $i = 1, \ldots, m \leq n$ the processes denote the dynamics of the basis assets, whereas for $i = m+1, \ldots, n$ the processes represent observable, but non-traded state variables. Moreover, we will return to our assumption of the existence of a deterministic MMA as a basis asset.

3.2.1 Diffusion Model

In a diffusion setting the sources of risk are modeled by an n-dimensional standard Brownian motion $W = (W^{(1)}, \ldots, W^{(n)})'$, having zero initial value and zero correlation. We assume that the Brownian motion generates the filtration. The dynamics of the adapted stochastic processes $S^{(i)}, i = 1, \ldots, n$, are given by the following stochastic differential equation (henceforth SDE):

$$dS_t^{(i)} = S_t^{(i)} \left(\mu_t^{(i)} dt + \sum_{j=1}^{n} \sigma_t^{(ij)} dW_t^{(j)} \right), \tag{3.1}$$

where $\mu_t^{(i)}$ is a predictable[1] process denoting the drift, and $\Sigma_t = (\sigma_t^{(ij)})$ is a root of the covariance matrix[2] with predictable processes as entries satisfying the common integrability conditions (see, e.g. Protter [36]). Recall that we only consider processes adapted to the filtration generated by the Brownian motion.

For the asset $S^{(i)}, i = 1, \ldots, m$, the drift $\mu_t^{(i)}$ can be interpreted as the expected return of the asset over an infinitesimal time interval $[t, t + dt]$. The dynamics of the MMA are given by

$$dB_t = r_t B_t dt,$$

with a deterministic interest rate r_t and $B_0 = 1$ so that $B_t = \exp\left(\int_0^t r_s ds\right)$.

3.2.2 Discrete Model

Without loss of generality we consider time steps of length $\Delta t = 1$ for the discrete setup. The basic input to describe the evolution of the underlying processes in a general discrete model is a vector of so-called basis martingales, denoted by $Z = (Z^{(1)}, \ldots, Z^{(n)})'$, representing the sources of risk. The basis martingales are assumed to generate the filtration. They are the discrete analogue to the n-dimensional Brownian motion W of 3.2.1. We assume that, given $\{(Z_u^{(1)}, \ldots, Z_u^{(n)})'; u = 0, 1, \ldots, t\}$, there are at most $n + 1$ realizations of the vector $(Z_{t+1}^{(1)}, \ldots, Z_{t+1}^{(n)})'$. This means that we can think of a (in general not recombining) tree model with at most $n + 1$ successor states for every node. In general, there can be redundancies in the increments of the basis martingales in some states and at some points in time, when the tree is degenerated, and there are less than $n + 1$ successor states at some point in time. In the non-degenerated case the increments of the basis martingales are assumed to be orthogonal conditional on $\mathcal{F}_t$. This means that the increments are conditionally uncorrelated with a conditional variance of one. In general, the basis martingales must satisfy the following conditions:

$$\mathbf{E}_t^P\left[\Delta Z_{t+1}^{(j)}\right] = 0, \quad \text{where } \Delta Z_{t+1}^{(j)} = Z_{t+1}^{(j)} - Z_t^{(j)}, j = 1, \ldots, n \tag{3.2}$$

$$\mathbf{E}_t^P\left[\Delta Z_{t+1}^{(j)} \Delta Z_{t+1}^{(i)}\right] = \begin{cases} 0 & i \neq j, \\ \{0, 1\} & i = j\,. \end{cases} \tag{3.3}$$

[1] For a definition of predictability in continuous time see, e.g., Protter [36].

[2] Note that since the covariance matrix is defined by $\Sigma\Sigma'$ the root is not unique.

Note that $\mathbf{E}_t^P\left[(\Delta Z_{t+1}^{(j)})^2\right] = 0$ implies $\Delta Z_{t+1}^{(j)} \equiv 0$ representing the degenerated case.

Again, we only consider processes adapted to the filtration generated by the basis martingales. The dynamics of the adapted processes $S^{(i)}, i = 1, \ldots, n$, are in analogy to the SDE (3.1) given by

$$S_t^{(i)} = S_{t-1}^{(i)}\left(1 + \mu_t^{(i)} + \sum_{j=1}^{n} \beta_t^{(ij)} \Delta Z_t^{(j)}\right), \tag{3.4}$$

where μ_t is predictable, i.e. adapted to $\mathcal{F}_{t-1}$, and $\beta_t = (\beta_t^{(ij)})$ is a root of the covariance matrix consisting of predictable entries with respect to the filtration. The representation in (3.4) follows from Doob's theorem which says that every adapted process has a unique decomposition into a predictable process and a martingale. The representation of martingales follows from the martingale representation theorem which states that every martingale $\mathcal{M}$ can be expressed as a linear combination of the basis martingales, i.e.

$$\mathcal{M}_t = \mathcal{M}_0 + \sum_{u=1}^{t}\sum_{j=1}^{n} \delta_u^{(j)} \Delta Z_u^{(j)},$$

with a predictable process δ. The definitions given above are according to Dothan [16]. The MMA follows the dynamics

$$B_t = B_{t-1}(1 + r_t),$$

where r_t denotes the deterministic short-term interest rate from $t-1$ to t, and $B_0 = 1$, so that $B_t = \prod_{s=1}^{t}(1 + r_s)$.

3.3 Model-Specific Change of Measure

We have seen in Chap. 2 that martingale pricing is a key concept in the framework of no-arbitrage pricing. In this context we have pointed out that a clever choice of numeraire allows for an elegant derivation of the pricing equation for a contingent claim. We now discuss the technique of changing the measure, explained in Chap.2, in the continuous and the discrete setup and apply it to the price processes, in order to derive the dynamics under the new measure. In the diffusion setup the results are well-known: When changing the measure, only the drift changes; the covariance remains the same.

In the discrete model we describe the change of measure along the lines of Dothan [16]. Moreover, we are able to extend his approach in proposing a definition of new basis martingales under the new measure in analogy to the new Brownian motion. Then, the aim is to write the dynamics under the new measure using the new basis martingales. We will see that apart from some minor restrictions this concept works in the discrete setup as well. We observe that in contrast to the diffusion model, the covariance structure of the basis martingales under the new measure changes in the discrete model. This implies that in the discrete setup higher moments play a decisive role.

3.3.1 Diffusion Model

In a diffusion model the change of measure between any two equivalent measures follows the theorem of Girsanov (see, e.g. Musiela and Rutkowski [33]).

Proposition 3.1 (Girsanov's Theorem). *Let the filtration $\mathcal{F}$ be generated by an n-dimensional Brownian motion W under P. Every measure $\bar{P}$ equivalent to P on $\mathcal{F}_T$ is of the form*

$$d\bar{P} = \exp\Big(- \int_0^T \sum_{j=1}^n \alpha_s^{(j)} dW_s^{(j)} - \frac{1}{2} \int_0^T \sum_{j=1}^n (\alpha_s^{(j)})^2 ds \Big) dP, \tag{3.5}$$

where $\alpha = (\alpha^{(1)}, \dots, \alpha^{(n)})'$ denotes an n-dimensional predictable process satisfying the common integrability conditions.[3] *Then, the process $\bar{W} = (W^{(1)}, \dots, W^{(n)})'$ defined by*

$$\bar{W}_t^{(j)} = W_t^{(j)} + \int_0^t \alpha_u^{(j)} du, \qquad j = 1, \dots, n, \tag{3.6}$$

is a standard n-dimensional Brownian motion under $\bar{P}$.

The converse implication is also true: For every n-dimensional predictable process α, Equation (3.5) defines a probability measure $\bar{P}$, equivalent to the physical measure P, since the density is indeed positive with an expectation of one assuming that the Novikov condition holds.[4]

Using this result, we are able to derive the dynamics of the stochastic processes under the new measure.

[3] These can again be found in Musiela [33].

[4] This is a sufficient condition for the exponential to be indeed a martingale, not only a local martingale (see, e.g. Baxter [4]).

Proposition 3.2 (Dynamics Under the New Measure). *For every stochastic process S following the SDE*

$$dS_t = S_t\mu_t dt + S_t \sum_{j=1}^{n} \sigma_t^{(j)} dW_t^{(j)},$$

the dynamics under the measure $\bar{P}$, given by α in Proposition 3.1, are

$$dS_t = S_t\left(\mu_t - \sum_{j=1}^{n} \alpha_t^{(j)}\sigma_t^{(j)}\right)dt + S_t \sum_{j=1}^{n} \sigma_t^{(j)} d\bar{W}_t^{(j)}, \tag{3.7}$$

where

$$d\bar{W}_t^{(j)} = dW_t^{(j)} + \alpha_t^{(j)} dt, \qquad j = 1, \ldots, n. \tag{3.8}$$

We see that the dynamics of the stochastic processes under any two equivalent measures only differ with respect to the drift term, whereas the covariance structure remains the same, since the covariance structure of the new Brownian motion has not changed. This result holds for any stochastic process, no matter if it represents a basis asset or a state variable.

3.3.2 Discrete Model

Similarly to the diffusion setup we can define a change of measure in the discrete setting: According to 2.3.1, any change of equivalent measures can be described by a martingale ξ. Using the martingale representation theorem and the martingale property of ξ, we can express ξ_T as

$$\begin{aligned}\xi_T &= \xi_{T-1}\left(1 - \sum_{j=1}^{n} \alpha_T^{(j)} \Delta Z_T^{(j)}\right) \\ &= \xi_0\left(1 - \sum_{j=1}^{n} \alpha_1^{(j)} \Delta Z_1^{(j)}\right) \cdots \left(1 - \sum_{j=1}^{n} \alpha_T^{(j)} \Delta Z_T^{(j)}\right),\end{aligned}$$

where $\xi_0 = 1$ and $\alpha = (\alpha^{(1)}, \ldots, \alpha^{(n)})'$ denotes an n-dimensional predictable process. With this representation we have derived a formula for the change between equivalent measures analogously to the diffusion setup: Every measure $\bar{P}$ equivalent to P is given by the following density:

$$d\bar{P} = \prod_{u=1}^{T}\left(1 - \sum_{j=1}^{n} \alpha_u^{(j)} \Delta Z_u^{(j)}\right) dP. \tag{3.9}$$

Equation (3.9) can be interpreted as a discrete approximation of formula (3.5); it is essentially the same as the one proposed by Dothan [16]. We can extend the analogy to the continuous setting even further by trying to find a set of new basis martingales with respect to the new measure $\bar{P}$. It is also possible to derive a representation of the dynamics of the stochastic processes with respect to the new measure using the new basis martingales, although imposing certain restrictions. We propose the following definition for a vector of new basis martingales $\bar{Z} = (\bar{Z}^{(1)}, \ldots, \bar{Z}^{(n)})'$ with respect to $\bar{P}$, setting

$$\Delta \bar{Z}_t^{(j)} := \Delta Z_t^{(j)} + \alpha_t^{(j)} \text{ for } \Delta Z_t^{(j)} \neq 0, \qquad j = 1, \ldots, n,$$

with $\bar{Z}_0^{(j)} = 0$ for $j = 1, \ldots, n$. This yields

$$\bar{Z}_t^{(j)} = Z_t^{(j)} + \sum_{u=1}^{t} \alpha_u^{(j)}, \qquad j = 1, \ldots, n. \tag{3.10}$$

If $\Delta Z_t^{(j)} \equiv 0$ for a component $Z^{(j)}$ we set $\Delta \bar{Z}_t^{(j)} \equiv 0$ as well. From now on we will only deal with the non-degenerated case to avoid case differentiation. Cum grano salis, the results also hold for the degenerated scenario. We see that formulae (3.9) and (3.10) are nothing but the discrete analogues to formulae (3.5) and (3.6). Essentially, we have derived a perfect analogy to the diffusion setup in proposing a change of measure that leads to new basis martingales. Nevertheless, there are two important differences to notice. The first concerns the change of measure: Using the change-of-measure formula, i.e. (3.5), in the diffusion setup, we obtain probability measures for all choices of α in (3.5) when the Novikov condition is satisfied. This is different in the discrete setup. The change of measure given by (3.9) guarantees for a measure normalized to one, according to the martingale properties of ξ. However, it can lead to so-called signed measures, i.e. measures which assign a negative value to an event for some choices of α. Thus, not all measures achieved by formula (3.9) are equivalent probability measures.[5]

The second difference concerns the properties of the Brownian motion and the basis martingales, respectively. In a diffusion model, the Brownian motion W is a P-martingale and the Brownian motion $\bar{W}$ a $\bar{P}$-martingale with identical second moments. In the discrete setup, $\bar{Z}$ is a $\bar{P}$-martingale as well,

[5] One could also take an exponential form of the density in (3.9 to avoid signed measures. However, this in turn would lead to further restrictions on the definition of the new basis martingales $\bar{Z}$.

however with second moments different from those of Z under P. Without loss of generality we take the process $\bar{Z}^{(i)}$ to prove that the components of $\bar{Z}$ are indeed martingales, i.e. that formula (3.2) is valid for $\bar{Z}$ under $\bar{P}$. We find

$$\begin{aligned}
\mathbf{E}_t^{\bar{P}}\left[\Delta\bar{Z}_{t+1}^{(i)}\right] &= \mathbf{E}_t\left[\frac{\xi_{t+1}}{\xi_t}\left(\Delta Z_{t+1}^{(i)}+\alpha_{t+1}^{(i)}\right)\right] \\
&= \mathbf{E}_t\left[\left(1-\sum_{j=1}^{n}\alpha_{t+1}^{(j)}\Delta Z_{t+1}^{(j)}\right)\left(\Delta Z_{t+1}^{(i)}+\alpha_{t+1}^{(i)}\right)\right] \\
&= 0-\alpha_{t+1}^{(i)}+\alpha_{t+1}^{(i)}+0=0.
\end{aligned}$$

Iterating this we get the martingale property. However, it is generally no longer true that the increments of the new martingales $\bar{Z}^{(1)},\ldots,\bar{Z}^{(n)}$ conditional on $\mathcal{F}_t$ are orthogonal. To be more precise, the new basis martingales under $\bar{P}$ do not have the following two properties that the original basis martingales have under the physical measure: Firstly, the increments of the new basis martingales have a conditional variance different from one. Secondly, they have non-zero conditional covariance, i.e. formula (3.3) is no longer valid. To prove this, we look at the following expression:

$$\begin{aligned}
\mathbf{E}_t^{\bar{P}}\left[\Delta\bar{Z}_{t+1}^{(i)}\Delta\bar{Z}_{t+1}^{(j)}\right] &= \mathbf{E}_t\Bigg[\left(1-\sum_{k=1}^{n}\alpha_{t+1}^{(k)}\Delta Z_{t+1}^{(k)}\right)\left(\Delta Z_{t+1}^{(i)}+\alpha_{t+1}^{(i)}\right)\times \\
&\qquad \left(\Delta Z_{t+1}^{(j)}+\alpha_{t+1}^{(j)}\right)\Bigg] \\
&= \mathbf{E}_t\Bigg[\Bigg(\Delta Z_{t+1}^{(i)}-\sum_{k=1}^{n}\alpha_{t+1}^{(k)}\Delta Z_{t+1}^{(k)}\Delta Z_{t+1}^{(i)}+\alpha_{t+1}^{(i)} \\
&\qquad -\alpha_{t+1}^{(i)}\sum_{k=1}^{n}\alpha_{t+1}^{(k)}\Delta Z_{t+1}^{(k)}\Bigg)\left(\Delta Z_{t+1}^{(j)}+\alpha_{t+1}^{(j)}\right)\Bigg] \\
&= \delta_{ij}-\sum_{k=1}^{n}\alpha_{t+1}^{(k)}\mathbf{E}_t\left[\Delta Z_{t+1}^{(k)}\Delta Z_{t+1}^{(i)}\Delta Z_{t+1}^{(j)}\right]-\alpha_{t+1}^{(i)}\alpha_{t+1}^{(j)},
\end{aligned}$$

where δ_{ij} denotes the Kronecker delta, which is one for $i=j$ and zero otherwise. Thus, higher moments of ΔZ appear in the covariance structure of the new basis martingales. This is different from the diffusion model: The new Brownian motion under $\bar{P}$ has the same features as the original one under the physical measure, which in particular means that the components of the new Brownian motion $\bar{W}$ are again uncorrelated under $\bar{P}$. We summarize our findings in:

Proposition 3.3 (Change of Measure in the Discrete Model). *Let the filtration $\mathcal{F}$ be generated by the basis martingales. Every measure $\bar{P}$ equivalent to P on $\mathcal{F}_T$ is of the form*

$$d\bar{P} = \prod_{u=1}^{T} \left(1 - \sum_{j=1}^{n} \alpha_u^{(j)} \Delta Z_u^{(j)}\right) dP,$$

where $\alpha = (\alpha^{(1)}, \ldots, \alpha^{(n)})'$ denotes an n-dimensional predictable process. There exists a vector of new basis martingales $\bar{Z} = (\bar{Z}^{(1)}, \ldots, \bar{Z}^{(n)})'$ with respect to $\bar{P}$ defined by

$$\Delta \bar{Z}_t^{(j)} = \Delta Z_t^{(j)} + \alpha_t^{(j)}, \qquad j = 1, \ldots, n,$$

where $\bar{Z}_0 = 0$ yielding

$$\bar{Z}_t^{(j)} = Z_t^{(j)} + \sum_{u=1}^{t} \alpha_u^{(j)}, \qquad j = 1, \ldots, n.$$

The increments of $\bar{Z}$ are no longer orthogonal conditional on $\mathcal{F}_t$.

Armed with this, we can easily calculate the dynamics of the stochastic processes under the new measure analogously to the diffusion setup.

Proposition 3.4 (Dynamics Under the New Measure). *For every stochastic process S following the dynamics*

$$S_t = S_{t-1} \left(1 + \mu_t + \sum_{j=1}^{n} \beta_t^{(j)} \Delta Z_t^{(j)}\right),$$

the dynamics under $\bar{P}$, given by α in Proposition 3.3, are

$$S_t = S_{t-1} \left(1 + \mu_t - \sum_{j=1}^{n} \alpha_t^{(j)} \beta_t^{(j)}\right) + S_{t-1} \sum_{j=1}^{n} \beta_t^{(j)} \Delta \bar{Z}_t^{(j)}, \tag{3.11}$$

where

$$\Delta \bar{Z}_t^{(j)} = \Delta Z_t^{(j)} + \alpha_t^{(j)}, \qquad j = 1, \ldots, n.$$

The dynamics under the physical measure P and the new measure $\bar{P}$ in formula (3.11) *seem* only to differ in the drift. However, this is not the case. Since the new basis martingales have a different covariance structure, the covariance structure of the stochastic processes $S^{(i)}, i = 1, \ldots, n$, under the measure $\bar{P}$ also has changed, compared with the covariance matrix under P, which is simply given by $\beta\beta'$. This result again holds for any stochastic process being a price process of an asset or a process of a state variable.

3.4 Normalized Price Processes

As we have seen in Chap. 2, numeraires play an important role in martingale pricing. The relevant numeraires in the derivation of the BS formula are the MMA and the underlying asset. Besides these, we take a look at an arbitrary positive portfolio as the numeraire consisting of a linear combination of the basis assets. This is of interest with regard to the power option in Chap. 4. We use these different numeraires to compute normalized price processes and apply the change of measure discussed in 3.3.1 and 3.3.2 within both the diffusion and the discrete setup. Then, we calculate the dynamics of the assets under the new measure, which is necessary for the explicit computation of the artificial probabilities as discussed in Chap. 2.

In contrast to Sect. 3.3, we only look at normalized *price* processes, and not at processes of the state variables, since the martingale property is valid only for normalized assets. However, having defined the change of measure such that the normalized assets are martingales under the corresponding measure, we can of course apply the change of measure to *all* exogenously given stochastic processes. This in particular includes the dynamics of the state variables, which have been derived for an arbitrary change of measure in Sect. 3.3. The drift of the processes representing state variables under the new measure is structurally different from the drift of the assets.

Since the calculations are rather involved, we provide an outline of the main findings. In the diffusion model, normalized price processes are diffusion processes as well, according to Itô's formula. The coefficients are straightforward to determine. In the discrete setup, there also exists a representation for normalized price processes according to the martingale representation theorem, but in general, the coefficients cannot be computed explicitly. The more complex the numeraire the more the structure of the dynamics will change. Whereas in the diffusion setup the changes of measure are concise, it is in general difficult to write down the change of measure explicitly in the discrete setup. Only for the deterministic numeraire, the MMA, are the drift restriction and the market prices of risk that determine the risk-neutral measures similar in both setups. Taking risky numeraires, the findings are different: As already shown for an arbitrary change of measure, the covariance structure does not change in the diffusion setup. Moreover, there is a convenient formula for the drift under every EMM corresponding to an arbitrary numeraire: It is simply equal to the sum of the short rate of interest and the instantaneous covariation of the asset return and the return of the numeraire. In the discrete setup, the covariance does not remain the same, and no compact formula for the drift exists for risky numeraires.

3.4.1 Discounted Price Processes and Risk-Neutral Measure

As defined in Chap. 2, the risk-neutral measures result from taking the MMA as the numeraire which means that discounted assets must be $\hat{P}$-martingales. Computing the discounted asset price processes and changing the measure such that the discounted processes have zero drift yields the drift restriction. This change of measure is induced by a process which can be interpreted as a market price of risk vector. Moreover, we are now able to understand the notion of risk-neutral pricing: The drift of the assets under the changed measure is equal to the risk-free rate earned by the MMA. Therefore, $\hat{P}$ is called a risk-neutral measure. Furthermore, martingale pricing using a risk-neutral measure is referred to as risk-neutral valuation.

3.4.1.1 Diffusion Model

We now take the MMA as the numeraire and compute the discounted price processes. We normalize the assets $S^{(i)}, i = 1, \ldots, m$, not the state variables, since only the assets must earn the risk-free rate under $\hat{P}$. Of course, we can derive the risk-neutral dynamics also for the non-traded state variables, similar to 3.3.1.

The dynamics of the discounted price processes are given by

$$d\left(\frac{S_t^{(i)}}{B_t}\right) = \frac{S_t^{(i)}}{B_t}\left((\mu_t^{(i)} - r_t)dt + \sum_{j=1}^{n} \sigma_t^{(ij)} dW_t^{(j)}\right), \qquad i = 1, \ldots, m,$$

using Itô's formula. The assumption of no-arbitrage yields the existence of an EMM, and in particular the existence of a measure $\hat{P}$ corresponding to the MMA as the numeraire. Thus, we can use Girsanov's theorem to eliminate the drift coefficient which gives

$$\begin{aligned} d\left(\frac{S_t^{(i)}}{B_t}\right) &= \frac{S_t^{(i)}}{B_t}\left((\mu_t^{(i)} - r_t)dt + \sum_{j=1}^{n} \sigma_t^{(ij)} dW_t^{(j)}\right) \\ &\overset{!}{=} \frac{S_t^{(i)}}{B_t} \sum_{j=1}^{n} \sigma_t^{(ij)} \left(dW_t^{(j)} + \lambda_t^{(j)} dt\right) = \frac{S_t^{(i)}}{B_t} \sum_{j=1}^{n} \sigma_t^{(ij)} d\hat{W}_t^{(j)} \end{aligned}$$

for a predictable n-dimensional process $\lambda = (\lambda^{(1)}, \ldots, \lambda^{(n)})'$.

$$\hat{W}_t^{(j)} = W_t^{(j)} + \int_0^t \lambda_s^{(j)} ds, \qquad j = 1, \ldots, n,$$

is a standard Brownian motion under $\hat{P}$. Thus, the following relationship must hold:

$$\mu_t^{(i)} = r_t + \sum_{j=1}^{n} \sigma_t^{(ij)} \lambda_t^{(j)}, \qquad i = 1, \ldots, m. \tag{3.12}$$

This is the so-called drift restriction, where λ can be interpreted as a vector of market prices of risk. The intuitive meaning of (3.12) is the following: The expected return is equal to the risk-free interest rate plus a risk premium which equals the amounts of risk, measured by the product of a root of the covariance and the respective market prices of risk. Thus, under the assumption of no-arbitrage there exists a market price of risk for each risk factor giving the compensation per unit of risk. This implies that for the same risk (expressed in terms of $dW^{(i)}$) the same compensation must be paid. Equation (3.12) in particular shows that each risk-neutral measure is determined by a density with respect to the physical measure using a market price of risk vector. The assumption of no-arbitrage does not exclude negative[6] market prices of risk, since only the existence of a process λ satisfying the drift restriction is assumed.

If the basis assets form a complete market, there are as many basis assets as sources of risk (i.e. $m = n$). This implies that the market prices of risk are uniquely determined by the drift, the volatility coefficients and the interest rate, since the drift restriction (3.12) can uniquely be solved for λ. In this case the risk-neutral measure is unique as well. In an incomplete market (i.e. $m < n$) there remain degrees of freedom for the market prices of risk. To eliminate these would require additional information such as additional price processes not carrying new sources of risk. Furthermore, the specific form of λ could be derived within an equilibrium model including a specification of investor's preferences (see, e.g. Cox, Ingersoll, and Ross [11]). Each vector λ satisfying (3.12) corresponds to a certain risk-neutral measure and vice versa.

[6] A negative market price of risk can be intuitive from an economic point of view when we for instance think of a risk-loving representative investor in an equilibrium setup.

Finally, changing the measure to derive the dynamics of the assets using the drift restriction, we end up with the drift equal to the risk-free rate under the corresponding risk-neutral measure. Thus we have derived:

Proposition 3.5 (Drift Restriction in the Diffusion Model). *Under the assumption of no-arbitrage there exists an n-dimensional predictable process λ such that for every asset S, following the SDE*

$$dS_t = S_t \mu_t dt + S_t \sum_{j=1}^{n} \sigma_t^{(j)} dW_t^{(j)},$$

the drift restriction

$$\mu_t = r_t + \sum_{j=1}^{n} \sigma_t^{(j)} \lambda_t^{(j)}$$

holds. $\lambda^{(j)}$ is a market price of risk for the risk factor $W^{(j)}$. The dynamics of S under the risk-neutral measure $\hat{P}$ – which is defined by the market price of risk vector λ and corresponds to the MMA as the numeraire – are

$$dS_t = S_t r_t dt + S_t \sum_{j=1}^{n} \sigma_t^{(j)} d\hat{W}_t^{(j)},$$

where

$$d\hat{W}_t^{(j)} = dW_t^{(j)} + \lambda_t^{(j)} dt.$$

$\hat{W}$ is a standard Brownian motion under the risk-neutral measure $\hat{P}$. The dynamics of the state variables under $\hat{P}$ are given by (3.7), replacing α by λ and $\bar{W}$ by $\hat{W}$ in (3.7) and (3.8).

3.4.1.2 Discrete Model

In the discrete model we can proceed analogously to the diffusion model. Taking the MMA as the numeraire, the discounted price processes of the assets $S^{(i)}, i = 1, \ldots, m$, are calculated as

$$\begin{aligned} \frac{S_t^{(i)}}{B_t} &= \frac{S_{t-1}^{(i)}}{B_{t-1}} \frac{1 + \mu_t^{(i)} + \sum_{j=1}^{n} \beta_t^{(ij)} \Delta Z_t^{(j)}}{1 + r_t} \\ &= \frac{S_{t-1}^{(i)}}{B_{t-1}} \left(1 + \frac{\mu_t^{(i)} - r_t}{1 + r_t} + \sum_{j=1}^{n} \hat{\beta}_t^{(ij)} \Delta Z_t^{(j)} \right), \end{aligned} \tag{3.13}$$

where

$$\hat{\beta}_t := \frac{\beta_t}{1+r_t} .$$

In order to obtain a representation for the discounted processes $S_t^{(i)}/B_t$ as a sum of a predictable expected return and a martingale, the matrix β_t changes to $\beta_t/(1+r_t)$.

As pointed out in 3.4.1.1 under the assumption of no-arbitrage there exists a risk-neutral measure in any setup. To derive the change of measure, we must eliminate the expected net return similar to the diffusion model. Thus, the right-hand side of (3.13) has to be equal to

$$\frac{S_{t-1}^{(i)}}{B_{t-1}}\left(1+\sum_{j=1}^{n}\hat{\beta}_t^{(ij)}(\Delta Z_t^{(j)}+\lambda_t^{(j)})\right) = \frac{S_{t-1}^{(i)}}{B_{t-1}}\left(1+\sum_{j=1}^{n}\hat{\beta}_t^{(ij)}\Delta \hat{Z}_t^{(j)}\right) \quad (3.14)$$

for a predictable n-dimensional process $\lambda = (\lambda^{(1)},\ldots,\lambda^{(n)})'$. The new basis martingales are given by

$$\Delta\hat{Z}_t^{(j)} = \Delta Z_t^{(j)} + \lambda_t^{(j)}, \qquad j=1,\ldots,n.$$

Rearranging the terms from (3.13) and (3.14) gives

$$\mu_t^{(i)} - r_t = \sum_{j=1}^{n}\beta_t^{(ij)}\lambda_t^{(j)}, \qquad i=1,\ldots,m, \qquad (3.15)$$

which is the drift restriction for the discrete setup (analogously to (3.12) in the diffusion setup). The interpretation of the result is in exact accordance with the diffusion setup and therefore omitted.

Finally, applying the change of measure to the dynamics of the assets and using the drift restriction (3.15) leads to the net return of r_t for every asset under the corresponding risk-neutral measure. Thus, the results are summarized in:

Proposition 3.6 (Drift Restriction in the Discrete Setup). *Under the assumption of no-arbitrage there exists an n-dimensional predictable process λ such that for every asset S, following the dynamics*

$$S_t = S_{t-1}\left(1+\mu_t+\sum_{j=1}^{n}\beta_t^{(j)}\Delta Z_t^{(j)}\right),$$

the drift restriction

$$\mu_t = r_t + \sum_{j=1}^{n} \beta_t^{(j)} \lambda_t^{(j)} \tag{3.16}$$

holds. $\lambda^{(j)}$ *is a market price of risk for the risk factor* $Z^{(j)}$*. The dynamics of* S *under the risk-neutral measure* $\hat{P}$ *– which is defined by the market price of risk vector* λ *and corresponds to the MMA as the numeraire – are*

$$S_t = S_{t-1}\left(1 + r_t + \sum_{j=1}^{n} \beta_t^{(j)} \Delta\hat{Z}_t^{(j)}\right),$$

where

$$\Delta\hat{Z}_t^{(j)} = \Delta Z_t^{(j)} + \lambda_t^{(j)}, \qquad j = 1, \ldots, n.$$

$\hat{Z}$ *is a vector of basis martingales under the risk-neutral measure* $\hat{P}$*. The dynamics of the state variables under* $\hat{P}$ *are given by (3.11), replacing* α *by* λ *and* $\bar{Z}$ *by* $\hat{Z}$ *in (3.10) and (3.11).*

3.4.2 Price Processes Normalized by a Risky Basis Asset

We now consider a risky numeraire, namely one of the basis assets. From a technical point of view, things become more complicated in this case, but at least in the diffusion setup there still remains a simple structure for the change of measure. In the discrete setup there is no simple explicit form for the variables determining the change of measure.

We now proceed as in 3.4.1. We take the asset $S^{(k)}$ as the numeraire and calculate the normalized price processes of the assets $S^{(i)}, i = 1, \ldots, m, i \neq k$. Then, under the assumption of no-arbitrage, we can change the measure such that the normalized price processes have zero drift. Finally, we compute the dynamics under the new measure.

3.4.2.1 Diffusion Model

When taking $S^{(k)}$ as the numeraire, the normalized price processes are calculated with Itô's formula, where $dS_t^{(i)} \bullet dS_t^{(j)}$ denotes the instantaneous covariation between $S^{(i)}$ and $S^{(j)}$:

$$
\begin{aligned}
d\left(\frac{S_t^{(i)}}{S_t^{(k)}}\right) &= 1/S_t^{(k)} dS_t^{(i)} + S_t^{(i)} d(1/S_t^{(k)}) + dS_t^{(i)} \bullet d(1/S_t^{(k)}) \\
&= \frac{S_t^{(i)}}{S_t^{(k)}}\left(\mu_t^{(i)} dt + \sum_{j=1}^{n} \sigma_t^{(ij)} dW_t^{(j)}\right) + \frac{S_t^{(i)}}{S_t^{(k)}}\left(-\mu_t^{(k)} + \sum_{j=1}^{n} (\sigma_t^{(kj)})^2\right) dt \\
&\quad - \frac{S_t^{(i)}}{S_t^{(k)}} \sum_{j=1}^{n} \sigma_t^{(kj)} dW_t^{(j)} - \frac{S_t^{(i)}}{S_t^{(k)}} \sum_{j=1}^{n}\sum_{l=1}^{n} \left(\sigma_t^{(kj)} \sigma_t^{(il)} dW_t^{(j)} \bullet dW_t^{(l)}\right),
\end{aligned}
$$

using

$$
d(1/S_t^{(k)}) = 1/S_t^{(k)}\left(-\mu_t^{(k)} + \sum_{j=1}^{n} (\sigma_t^{kj})^2\right) dt + 1/S_t^{(k)} \sum_{j=1}^{n} (-\sigma_t^{(kj)})\, dW_t^{(j)}.
$$

This yields

$$
\begin{aligned}
d\left(\frac{S_t^{(i)}}{S_t^{(k)}}\right) &= \frac{S_t^{(i)}}{S_t^{(k)}}\Bigg[\left(\mu_t^{(i)} - \mu_t^{(k)} + \sum_{j=1}^{n} \sigma_t^{(kj)} (\sigma_t^{(kj)} - \sigma_t^{(ij)})\right) dt \\
&\qquad + \sum_{j=1}^{n} (\sigma_t^{(ij)} - \sigma_t^{(kj)})\, dW_t^{(j)}\Bigg] \\
&\overset{!}{=} (S_t^{(i)}/S_t^{(k)}) \sum_{j=1}^{n} (\sigma_t^{(ij)} - \sigma_t^{(kj)}) \left(dW_t^{(j)} + \tilde{\lambda}_t^{(j)} dt\right) \\
&= (S_t^{(i)}/S_t^{(k)}) \sum_{j=1}^{n} (\sigma_t^{(ij)} - \sigma_t^{(kj)})\, d\tilde{W}_t^{(j)}
\end{aligned}
$$

under the assumption of no-arbitrage for a predictable n-dimensional process $\tilde{\lambda}$.

$$
\tilde{W}_t^{(j)} = W_t^{(j)} + \int_0^t \tilde{\lambda}_s^{(j)} ds, \quad j = 1, \ldots, n,
$$

is a standard Brownian motion under $\tilde{P}$, which is defined by $\tilde{\lambda}$ and corresponds to $S^{(k)}$ as the numeraire. Thus, the change of measure such that price processes normalized with $S_t^{(k)}$ are martingales is given by Girsanov's theorem with a predictable process $\tilde{\lambda}$ satisfying the following relation:

$$
\sum_{j=1}^{n} (\sigma_t^{(ij)} - \sigma_t^{(kj)}) \tilde{\lambda}_t^{(j)} = \mu_t^{(i)} - \mu_t^{(k)} + \sum_{j=1}^{n} \sigma_t^{(kj)} (\sigma_t^{(kj)} - \sigma_t^{(ij)}). \quad (3.17)
$$

Since the normalized MMA must be a martingale as well, it has to be true that

$$\sum_{j=1}^{n}(-\sigma_t^{(kj)})\tilde{\lambda}_t^{(j)} = r_t - \mu_t^{(k)} + \sum_{j=1}^{n}(\sigma_t^{(kj)})^2. \tag{3.18}$$

This result can be obtained by calculating $d(B_t/S_t^{(k)})$ and eliminating the drift. Subtracting (3.18) from (3.17) we arrive at the following drift restriction:

$$\mu_t^{(i)} = r_t + \sum_{j=1}^{n}\sigma_t^{(ij)}(\tilde{\lambda}_t^{(j)} + \sigma_t^{(kj)}). \tag{3.19}$$

In a complete market $\tilde{\lambda}$ is unique. In an incomplete market there exists an EMM $\tilde{P}$ for each $\tilde{\lambda}$ and vice versa. Note that formula (3.19) can be rewritten as

$$\tilde{\lambda}_t^{(j)} = \lambda_t^{(j)} - \sigma_t^{(kj)},$$

since there is a one-to-one correspondence between $\tilde{\lambda}$ and λ. This in particular shows that the change of measure from P to $\tilde{P}$ again involves the market prices of risk. We sum up the findings in:

Proposition 3.7 (Dynamics under the EMM with Asset $\mathbf{S^{(k)}}$ as Numeraire). *Under the assumption of no-arbitrage, there exists an n-dimensional predictable process $\tilde{\lambda}$ such that for every asset S, following the SDE*

$$dS_t = S_t\mu_t dt + S_t\sum_{j=1}^{n}\sigma_t^{(j)}dW_t^{(j)},$$

the drift restriction

$$\mu_t = r_t + \sum_{j=1}^{n}\sigma_t^{(j)}(\tilde{\lambda}_t^{(j)} + \sigma_t^{(kj)})$$

holds. The dynamics under the measure $\tilde{P}$ – which is defined by $\tilde{\lambda}$ and corresponds to the numeraire $S^{(k)}$ – are

$$\begin{aligned} dS_t &= S_t\left(r_t + \sum_{j=1}^{n}\sigma_t^{(j)}\sigma_t^{(kj)}\right)dt + S_t\sum_{j=1}^{n}\sigma_t^{(j)}d\tilde{W}_t^{(j)} \qquad (3.20)\\ &= S_t\left(r_t + \frac{dS_t}{S_t}\bullet\frac{dS_t^{(k)}}{S_t^{(k)}}\right)dt + S_t\sum_{j=1}^{n}\sigma_t^{(j)}d\tilde{W}_t^{(j)}, \end{aligned}$$

where

$$d\tilde{W}_t^{(j)} = dW_t^{(j)} + \tilde{\lambda}_t^{(j)}dt.$$

$\tilde{W}$ is a standard Brownian motion under $\tilde{P}$. The dynamics of the state variables under $\tilde{P}$ are given by (3.7), replacing α by $\tilde{\lambda}$ and $\bar{W}$ by $\tilde{W}$ in (3.7) and (3.8).

The features of the asset price dynamics under $\tilde{P}$ are the following: Firstly, due to 3.3.1 the covariance structure remains unchanged. Secondly, the drift under the new measure is equal to the sum of the short-term interest rate and the instantaneous covariation between the asset return dS/S and the return of the numeraire asset $dS^{(k)}/S^{(k)}$.

3.4.2.2 Discrete Model

As we have seen in 3.4.1.2 when using the MMA as the numeraire, the first step is to derive the martingale representation for the normalized processes. This is quite complicated for risky numeraires in the discrete setup. In principle, the technique based on normalized price processes still works, however, the explicit computation of the process describing the change of measure is tedious.

We begin with the computation of the martingale representation for normalized price processes. It must hold that

$$\begin{aligned} S_t^{(i)}/S_t^{(k)} &= S_{t-1}^{(i)}/S_{t-1}^{(k)} \frac{1+\mu_t^{(i)} + \sum_{j=1}^n \beta_t^{(ij)} \Delta Z_t^{(j)}}{1+\mu_t^{(k)} + \sum_{j=1}^n \beta_t^{(kj)} \Delta Z_t^{(j)}} \\ &\stackrel{!}{=} S_{t-1}^{(i)}/S_{t-1}^{(k)} \left(1+\tilde{\mu}_t^{(i)} + \sum_{j=1}^n \tilde{\beta}_t^{(ij)} \Delta Z_t^{(j)}\right) \end{aligned}$$

for all $i = 1, \ldots, m, k \neq i$. This is equivalent to

$$\begin{aligned} \mu_t^{(i)} + \sum_{j=1}^n \beta_t^{(ij)} \Delta Z_t^{(j)} &= \mu_t^{(k)} + (1+\mu_t^{(k)})\tilde{\mu}_t^{(i)} + \sum_{j=1}^n \sum_{l=1}^n \beta_t^{(kj)} \tilde{\beta}_t^{(il)} \Delta Z_t^{(j)} \Delta Z_t^{(l)} \\ &+ \sum_{j=1}^n \Delta Z_t^{(j)} \left((1+\tilde{\mu}_t^{(i)})\beta_t^{(kj)} + (1+\mu_t^{(k)})\tilde{\beta}_t^{(ij)}\right). \end{aligned} \tag{3.21}$$

To derive a system of equations to uniquely determine the $n+1$ unknowns $\tilde{\mu}_t^{(i)}$ and $\tilde{\beta}_t^{(ij)}, j = 1, \ldots, n$, for fixed $i \neq k$ we first apply the expectation operator to (3.21). Using the properties of the basis martingales we get the following equation:

$$\mu_t^{(i)} = \mu_t^{(k)} + (1+\mu_t^{(k)})\tilde{\mu}_t^{(i)} + \sum_{j=1}^{n} \beta_t^{(kj)} \tilde{\beta}_t^{(ij)}.$$

Second, we multiply (3.21) by $\Delta Z^{(u)}, u = 1, \dots, n$, and then take the expectations. This yields n additional equations for $u = 1, \dots, n$:

$$\beta_t^{(iu)} = (1+\tilde{\mu}_t^{(i)})\beta_t^{(ku)} + (1+\mu_t^{(k)})\tilde{\beta}_t^{(iu)} + \sum_{j=1}^{n}\sum_{l=1}^{n} \beta_t^{(kj)} \tilde{\beta}_t^{(il)} \mathbf{E}_{t-1}[\Delta Z_t^{(j)} \Delta Z_t^{(l)} \Delta Z_t^{(u)}].$$

We can now solve uniquely for $\tilde{\mu}_t^{(i)}, \tilde{\beta}_t^{(iu)}, u = 1, \dots, n$, for fixed $i \neq k$.

Moreover, an analogous equation has to hold for the MMA:

$$\begin{aligned} B_t/S_t^{(k)} &= B_{t-1}/S_{t-1}^{(k)} \frac{1+r_t}{1+\mu_t^{(k)} + \sum_{j=1}^{n} \beta_t^{(kj)} \Delta Z_t^{(j)}} \\ &\stackrel{!}{=} B_{t-1}/S_{t-1}^{(k)} \left(1 + \tilde{\mu}_t^B + \sum_{j=1}^{n} \tilde{\beta}_t^{(Bj)} \Delta Z_t^{(j)}\right). \end{aligned}$$

Having computed the coefficients $\tilde{\mu}^B$ and $\tilde{\beta}^B$, we can apply the change of measure discussed in 3.3.2 to achieve a zero net return under the new measure $\tilde{P}$, i.e.

$$\begin{aligned} S_t^{(i)}/S_t^{(k)} &\stackrel{!}{=} S_{t-1}^{(i)}/S_{t-1}^{(k)} \left(1 + \sum_{j=1}^{n} \tilde{\beta}_t^{(ij)} \tilde{\Delta} Z_t^{(j)}\right) \\ &= S_{t-1}^{(i)}/S_{t-1}^{(k)} \left(1 + \sum_{j=1}^{n} \tilde{\beta}_t^{(ij)} (\Delta Z_t^{(j)} + \tilde{\lambda}_t^{(j)})\right), \\ B_t/S_t^{(k)} &\stackrel{!}{=} B_{t-1}/S_{t-1}^{(k)} \left(1 + \sum_{j=1}^{n} \tilde{\beta}_t^{(Bj)} \tilde{\Delta} Z_t^{(j)}\right) \\ &= B_{t-1}/S_{t-1}^{(k)} \left(1 + \sum_{j=1}^{n} \tilde{\beta}_t^{(Bj)} (\Delta Z_t^{(j)} + \tilde{\lambda}_t^{(j)})\right), \end{aligned}$$

under the assumption of no-arbitrage for a predictable n-dimensional process $\tilde{\lambda} = (\tilde{\lambda}^{(1)}, \dots, \tilde{\lambda}^{(n)})'$. The basis martingales under $\tilde{P}$ are given by

$$\Delta \tilde{Z}_t^{(j)} = \Delta Z_t^{(j)} + \tilde{\lambda}_t^{(j)}, \qquad j = 1, \dots, n,$$

where $\tilde{\lambda}_t$ has to satisfy the following n equations:

$$\tilde{\mu}_t^{(i)} = \sum_{j=1}^{n} \tilde{\beta}_t^{(ij)} \tilde{\lambda}_t^{(j)} \text{ for } k \neq i, \tag{3.22}$$

$$\tilde{\mu}_t^{B} = \sum_{j=1}^{n} \tilde{\beta}_t^{(Bj)} \tilde{\lambda}_t^{(j)}. \tag{3.23}$$

Plugging these relationships into the original dynamics, we end up with the dynamics under the new EMM. The results are recapitulated in:

Proposition 3.8 (Dynamics under the EMM with Asset $\mathbf{S^{(k)}}$ as Numeraire). *For every asset S, following the dynamics*

$$S_t = S_{t-1}\left(1 + \mu_t + \sum_{j=1}^{n} \beta_t^{(j)} \Delta Z_t^{(j)}\right),$$

there exist coefficients $\tilde{\mu}$ and $\tilde{\beta}$ such that the normalized price process is given by

$$S_t/S_t^{(k)} = S_{t-1}/S_{t-1}^{(k)}\left(1 + \tilde{\mu}_t + \sum_{j=1}^{n} \tilde{\beta}_t^{(j)} \Delta Z_t^{(j)}\right),$$

due to the martingale representation theorem. Furthermore, under the assumption of no-arbitrage there exists an n-dimensional predictable process $\tilde{\lambda}$ satisfying

$$\tilde{\mu}_t = \sum_{j=1}^{n} \tilde{\beta}_t^{(j)} \tilde{\lambda}_t^{(j)}. \tag{3.24}$$

The dynamics of the assets under the measure $\tilde{P}$ – which is defined by $\tilde{\lambda}$ and corresponds to the numeraire $S^{(k)}$ – are

$$S_t = S_{t-1}\left(1 + \mu_t - \sum_{j=1}^{n} \beta_t^{(j)} \tilde{\lambda}_t^{(j)}\right) + S_{t-1} \sum_{j=1}^{n} \beta_t^{(j)} \Delta \tilde{Z}_t^{(j)},$$

where

$$\Delta \tilde{Z}_t^{(j)} = \Delta Z_t^{(j)} + \tilde{\lambda}_t^{(j)}, \quad j = 1, \ldots, n.$$

$\tilde{Z}$ is a vector of basis martingales under $\tilde{P}$.

In contrast to the diffusion model we are not able to derive an explicit relation between the market prices of risk λ and the process $\tilde{\lambda}$ in the discrete setup. This is due to the fact that the coefficients of the normalized processes have no simple form, thus a representation in terms of economic variables is not available. In 3.5.1 we will deal with a simple setup consisting only of one risky asset and the MMA. There, we normalize the MMA with the risky asset in order to compute the coefficients and the change of measure explicitly. Even in this case things are more complicated than one might have expected.

3.4.3 Price Processes Normalized by a Portfolio

We now consider the most general numeraire feasible within our definition which is a positive, self-financing portfolio V consisting of the basis assets. As we have seen in Chap. 2, any change between EMM corresponds to a change of numeraires. We now discuss the dynamics under the EMM corresponding to any self-financing, positive portfolio. This in particular includes the results for a risky basis asset and the MMA as numeraires for appropriate choices of the coefficients in the numeraire portfolio. The proceeding is the same as in the previous subsections. We will see an application for a numeraire portfolio in Chap. 4 when computing the pricing equation for the power option in the BS setup.

3.4.3.1 Diffusion Model

We now build a self-financing portfolio consisting of the basis assets $S^{(i)}, i = 1, \ldots, m$, and the MMA B, setting

$$V_t = \sum_{k=1}^{m} x_t^{(k)} S_t^{(k)} + x_t^B B_t. \tag{3.25}$$

The numbers of units of the risky basis assets held at t are denoted by $x_t^{(i)}$, and x_t^B for the MMA. Since we want to establish a self-financing portfolio, not all coefficients $x_t^{(i)}, i = 1, \ldots, m$ and x_t^B can be chosen arbitrarily. Without loss of generality we use the MMA for the residual position to ensure that the self-financing property holds. This property can be expressed through the following equation (cf., inter alia, Björk [6]):

$$dV_t = \sum_{k=1}^{m} x_t^{(k)} dS_t^{(k)} + x_t^B r_t B_t dt.$$

With Itô's formula we obtain the following dynamics for the normalized assets, taking the portfolio as the numeraire:

$$\begin{aligned}
d\left(\frac{S_t^{(i)}}{V_t}\right) &= 1/V_t dS_t^{(i)} + S_t^{(i)} d(1/V_t) + dS_t^{(i)} \bullet d(1/V_t) \\
&= \frac{S_t^{(i)}}{V_t}\left(\mu_t^{(i)} - r_t \frac{x_t^B B_t}{V_t} - \sum_{k=1}^{m} \frac{x_t^{(k)} S^{(k)}}{V_t} \mu_t^{(k)} - \sum_{k=1}^{m} \frac{x_t^{(k)} S_t^{(k)}}{V_t} \times \right. \\
&\qquad \left. \sum_{j=1}^{n} \sigma_t^{(kj)} \sigma_t^{(ij)} + \sum_{k=1}^{m} \sum_{l=1}^{m} \frac{x_t^{(k)} S_t^{(k)}}{V_t} \frac{x_t^{(l)} S_t^{(l)}}{V_t} \sum_{j=1}^{n} \sigma_t^{(kj)} \sigma_t^{(lj)} \right) dt \\
&\quad + \frac{S_t^{(i)}}{V_t} \sum_{j=1}^{n} \left(\sigma_t^{(ij)} - \sum_{k=1}^{m} \frac{x_t^{(k)} S_t^{(k)}}{V_t} \sigma_t^{(kj)}\right) dW_t^{(j)},
\end{aligned}$$

using

$$\begin{aligned}
d(1/V_t) = &\left(- r_t \frac{x_t^B B_t}{V_t^2} - \sum_{k=1}^{m} \frac{x_t^{(k)} S_t^{(k)}}{V_t^2} \mu_t^{(k)} + \sum_{k=1}^{m} \sum_{l=1}^{m} \frac{x_t^{(k)} S_t^{(k)} x_t^{(l)} S_t^{(l)}}{V_t^3} \times \right. \\
&\left. \sum_{j=1}^{n} \sigma_t^{(kj)} \sigma_t^{(lj)} \right) dt - \sum_{k=1}^{m} \sum_{j=1}^{n} \frac{x_t^{(k)} S_t^{(k)}}{V_t^2} \sigma_t^{(kj)} dW_t^{(j)}.
\end{aligned}$$

Now, the relationship

$$-r_t \frac{x_t^B B_t}{V_t} - \sum_{k=1}^{m} \frac{x_t^{(k)} S_t^{(k)}}{V_t} \mu_t^{(k)} = -r_t + \sum_{k=1}^{m} \frac{x_t^{(k)} S_t^{(k)}}{V_t} (r_t - \mu_t^{(k)}),$$

which follows directly from the definition of the portfolio in (3.25), yields

$$\begin{aligned}
d\left(\frac{S_t^{(i)}}{V_t}\right) &= \frac{S_t^{(i)}}{V_t}\left(\mu_t^{(i)} - r_t + \sum_{k=1}^{m} \frac{x_t^{(k)} S_t^{(k)}}{V_t} (r_t - \mu_t^{(k)}) - \sum_{k=1}^{m} \frac{x_t^{(k)} S_t^{(k)}}{V_t} \times \right. \\
&\qquad \left. \sum_{j=1}^{n} \sigma_t^{(kj)} \sigma_t^{(ij)} + \sum_{k=1}^{n} \sum_{l=1}^{n} \frac{x_t^{(k)} S_t^{(k)}}{V_t} \frac{x_t^{(l)} S_t^{(l)}}{V_t} \sum_{j=1}^{n} \sigma_t^{(kj)} \sigma_t^{(lj)} \right) dt \\
&\quad + \frac{S_t^{(i)}}{V_t} \sum_{j=1}^{n} \left(\sigma_t^{(ij)} - \sum_{k=1}^{m} \frac{x_t^{(k)} S_t^{(k)}}{V_t} \sigma_t^{(kj)}\right) dW_t^{(j)} \\
&\overset{!}{=} (S_t^{(i)}/V_t) \sum_{j=1}^{n} \left(\sigma_t^{(ij)} - \sum_{k=1}^{m} \frac{x_t^{(k)} S_t^{(k)}}{V_t} \sigma_t^{(kj)}\right) \left(dW_t^{(j)} + \breve{\lambda}_t^{(j)} dt\right) \\
&= (S_t^{(i)}/V_t) \sum_{j=1}^{n} \left(\sigma_t^{(ij)} - \sum_{k=1}^{m} \frac{x_t^{(k)} S_t^{(k)}}{V_t} \sigma_t^{(kj)}\right) d\breve{W}_t^{(j)}
\end{aligned}$$

under the assumption of no-arbitrage for an n-dimensional predictable process $\breve{\lambda}$.

$$\breve{W}_t^{(j)} = W_t^{(j)} + \int_0^t \breve{\lambda}_s^{(j)} ds, \quad j = 1, \ldots, n,$$

is a standard Brownian motion under $\breve{P}$, which is defined by $\breve{\lambda}$ and corresponds to V as the numeraire. Thus, the change of measure is given by Girsanov's theorem with the predictable process $\breve{\lambda}$, satisfying

$$\sum_{j=1}^{n}\left(\sigma_t^{(ij)}-\sum_{k=1}^{m}\frac{x_t^{(k)}S_t^{(k)}}{V_t}\sigma_t^{(kj)}\right)\breve{\lambda}_t^{(j)}=\mu_t^{(i)}-r_t+\sum_{k=1}^{m}\frac{x_t^{(k)}S_t^{(k)}}{V_t}(r_t-\mu_t^{(k)})$$
$$-\sum_{k=1}^{m}\frac{x_t^{(k)}S_t^{(k)}}{V_t}\sum_{j=1}^{n}\sigma_t^{(kj)}\sigma_t^{(ij)}+\sum_{k=1}^{m}\frac{x_t^{(k)}S_t^{(k)}}{V_t}\sum_{l=1}^{m}\frac{x_t^{(l)}S_t^{(l)}}{V_t}\sum_{j=1}^{n}\sigma_t^{(kj)}\sigma_t^{(lj)}. \quad (3.26)$$

Since the normalized MMA must be a martingale as well, it has to hold that

$$\sum_{j=1}^{n}\left(-\sum_{k=1}^{m}\frac{x_t^{(k)}S_t^{(k)}}{V_t}\sigma_t^{(kj)}\right)\breve{\lambda}_t^{(j)}=\sum_{k=1}^{m}\frac{x_t^{(k)}S_t^{(k)}}{V_t}(r_t-\mu_t^{(k)})+\sum_{k=1}^{m}\frac{x_t^{(k)}S_t^{(k)}}{V_t}\times$$
$$\sum_{l=1}^{m}\frac{x_t^{(l)}S_t^{(l)}}{V_t}\sum_{j=1}^{n}\sigma_t^{(kj)}\sigma_t^{(lj)}. \quad (3.27)$$

This result is again achieved by calculating $d(B_t/V_t)$ and eliminating the drift. Subtracting (3.27) from (3.26) we find the following drift restriction using the portfolio as the numeraire:

$$\mu_t^{(i)}=r_t+\sum_{j=1}^{n}\sigma_t^{(ij)}\left(\breve{\lambda}_t^{(j)}+\sum_{k=1}^{m}\frac{x_t^{(k)}S_t^{(k)}}{V_t}\sigma_t^{(kj)}\right). \quad (3.28)$$

In perfect analogy to 3.4.2.1 the following findings are true in the continuous setting: Firstly, in a complete market $\breve{\lambda}$ is unique. Secondly, in an incomplete market there is a one-to-one correspondence between each EMM $\breve{P}$ and each $\breve{\lambda}$ and vice versa. Thirdly, Equation (3.28) can be rewritten as

$$\breve{\lambda}_t^{(j)}=\lambda_t^{(j)}-\sum_{k=1}^{m}\frac{x_t^{(k)}S_t^{(k)}}{V_t}\sigma_t^{(kj)},$$

which again allows us to describe the change of measure including the market prices of risk. We sum up our findings in:

Proposition 3.9 (Dynamics under the EMM with a Numeraire Portfolio). *Under the assumption of no-arbitrage there exists an n-dimensional predictable process $\breve{\lambda}$ such that for every asset S, following the SDE*

$$dS_t=S_t\mu_t dt+S_t\sum_{j=1}^{n}\sigma_t^{(j)}dW_t^{(j)},$$

the drift restriction

$$\mu_t = r_t + \sum_{j=1}^{n} \sigma_t^{(j)} \left(\breve{\lambda}_t^{(j)} + \sum_{k=1}^{m} \frac{x_t^{(k)} S_t^{(k)}}{V_t} \sigma_t^{(kj)} \right)$$

holds. The dynamics under the measure $\breve{P}$ – which is defined by $\breve{\lambda}$ and corresponds to the numeraire portfolio V – are

$$\begin{aligned} dS_t &= S_t \left(r_t + \sum_{j=1}^{n} \sigma_t^{(j)} \sum_{k=1}^{m} \frac{x_t^{(k)} S_t^{(k)}}{V_t} \sigma_t^{(kj)} \right) dt + S_t \sum_{j=1}^{n} \sigma_t^{(j)} d\breve{W}_t^{(j)} \qquad (3.29) \\ &= S_t \left(r_t dt + \frac{dS_t}{S_t} \bullet \frac{dV_t}{V_t} \right) + S_t \sum_{j=1}^{n} \sigma_t^{(j)} d\breve{W}_t^{(j)}, \end{aligned}$$

where

$$d\breve{W}_t^{(j)} = dW_t^{(j)} + \breve{\lambda}_t^{(j)} dt.$$

$\breve{W}$ is a standard Brownian motion under $\breve{P}$. The dynamics of the state variables under $\breve{P}$ are given by (3.7), replacing α by $\breve{\lambda}$ and $\bar{W}$ by $\breve{W}$ in (3.7) and (3.8).

The essential properties of the asset price dynamics under the measure $\breve{P}$ are the following: The drift has changed to the sum of the short-term interest rate and the instantaneous covariation between the asset return dS_t/S_t and the return of the numeraire portfolio dV_t/V_t. This formula is valid for all choices of V. The findings for the MMA as the numeraire derived in 3.4.1.1 and a risky basis asset considered in 3.4.2.1 are special cases of this result: Taking the MMA as the numeraire, we end up with the drift equal to the short-term interest rate, because the covariation is equal to zero since the MMA is deterministic. Taking a risky basis asset as the numeraire, the drift is equal to the interest rate plus the covariation of the underlying return and the return of the risky basis asset. Hence, we have proved the following:

Theorem 3.10. *In a diffusion model, the drift of the assets under any EMM is equal to the sum of the short-term interest rate and the instantaneous covariation between the asset return and the numeraire return. The covariance structure of the assets remains unchanged.*

3.4.3.2 Discrete Model

We can proceed analogously to 3.4.2.2. First, we must compute the martingale representation, i.e. the coefficients, for the normalized processes. Second, we have to derive the drift restriction for $\breve{\lambda}$ similar to (3.24). With the help of the drift restriction we can finally compute the dynamics under the new measure. Since the previous analysis has shown that the formulae are rather complicated and possibly no more explicit in a discrete setup, we refrain from the elaborate derivations. We summarize the results in:

Proposition 3.11 (Dynamics under the EMM with a Numeraire Portfolio). *Let V be a self-financing portfolio consisting of the basis assets. For every asset S, following the dynamics*

$$S_t = S_{t-1}\left(1 + \mu_t + \sum_{j=1}^{n} \beta_t^{(j)} \Delta Z_t^{(j)}\right),$$

there exist coefficients $\breve{\mu}$ and $\breve{\beta}$ such that the normalized price process is given by

$$S_t/V_t = S_{t-1}/V_{t-1}\left(1 + \breve{\mu}_t + \sum_{j=1}^{n} \breve{\beta}_t^{(j)} \Delta Z_t^{(j)}\right),$$

due to the martingale representation theorem. Furthermore, under the assumption of no-arbitrage there exists an n-dimensional predictable process $\breve{\lambda}$ satisfying

$$\breve{\mu}_t = \sum_{j=1}^{n} \breve{\beta}_t^{(j)} \breve{\lambda}_t^{(j)}. \tag{3.30}$$

The dynamics of the assets under the measure $\breve{P}$ – which is defined by $\breve{\lambda}$ and corresponds to the numeraire portfolio V – are

$$S_t = S_{t-1}\left(1 + \mu_t - \sum_{j=1}^{n} \beta_t^{(j)} \breve{\lambda}_t^{(j)}\right) + S_{t-1} \sum_{j=1}^{n} \beta_t^{(j)} \Delta \breve{Z}_t^{(j)},$$

where

$$\Delta \breve{Z}_t^{(j)} = \Delta Z_t^{(j)} + \breve{\lambda}_t^{(j)}, \quad j = 1, \dots, n.$$

$\breve{Z}$ is a vector of basis martingales under $\breve{P}$.

3.5 Examples

We begin with the discussion of two examples in the discrete setup to further illustrate the general results. First, we look at a risky asset as the numeraire in a simple model of a complete market with two basis assets only and compute the change of measure explicitly. For the common specification of a binomial tree we calculate the basis martingale and the density, which gives the risk-neutral measure, explicitly with the help of the market price of risk. Moreover, we discuss two examples in the diffusion model: Firstly, a complete market setup with two basis assets, and secondly, one basis asset with stochastic volatility serving as the most prominent example for an incomplete market. The objective is to compute the risk-neutral dynamics and to provide a detailed investigation of the market price of risk vector.

3.5.1 Complete Market with Two Basis Assets in the Discrete Setup

We consider the simplest scenario of a complete market consisting of one risky asset S and the MMA B. In this case we need only one basis martingale to describe the dynamics of the risky asset. This especially implies that there are only two successor states in each time step. Hence, this scenario in fact represents a simple binomial tree. Now, we calculate the coefficients of the MMA, normalized with the risky asset, in the martingale representation explicitly. This example shows that even in this simple setting the technique of normalizing price processes, being the standard approach in the continuous setup, gets rather complicated.

Starting from the equation

$$B_t/S_t = B_{t-1}/S_{t-1}\,\frac{1+r_t}{1+\mu_t+\beta_t\Delta Z_t} \stackrel{!}{=} B_{t-1}/S_{t-1}\left(1+\tilde{\mu}_t^B+\tilde{\beta}_t^B\Delta Z_t\right)$$

and proceeding as in 3.4.2.2, we obtain the following system of equations to determine $\tilde{\mu}_t^B$ and $\tilde{\beta}_t^B$:

$$\begin{aligned} r_t &= \mu_t + \tilde{\mu}_t^B + \mu_t\tilde{\mu}_t^B + \beta_t\tilde{\beta}_t^B, \\ 0 &= (1+\tilde{\mu}_t^B)\beta_t + (1+\mu_t)\tilde{\beta}_t^B + \beta_t\tilde{\beta}_t^B\mathbf{E}_{t-1}\left[(\Delta Z_t)^3\right]. \end{aligned}$$

Solving for the unknowns we get

$$\begin{aligned} \tilde{\beta}_t^B &= \frac{-\beta_t(1+r_t)}{(1+\mu_t)^2-\beta_t^2+\beta_t(1+\mu_t)\mathbf{E}_{t-1}\left[(\Delta Z_t)^3\right]} \\ \tilde{\mu}_t^B &= \frac{r_t-\mu_t}{1+\mu_t} + \frac{\beta_t^2(1+r_t)}{(1+\mu_t)^3-\beta_t^2(1+\mu_t)+\beta_t(1+\mu_t)^2\mathbf{E}_{t-1}\left[(\Delta Z_t)^3\right]}. \end{aligned}$$

The change of measure from P to $\tilde{P}$ is induced by

$$\tilde{\lambda}_t = \tilde{\mu}_t^B / \tilde{\beta}_t^B .$$

Then, B/S is a $\tilde{P}$-martingale. This yields the new dynamics

$$S_t = S_{t-1}\left(1 + \mu_t - \frac{\tilde{\mu}_t^B}{\tilde{\beta}_t^B}\beta_t + \beta_t \Delta \tilde{Z}_t\right) .$$

The expected return can be computed explicitly, plugging in the definition of $\tilde{\mu}_t, \tilde{\beta}_t$:

$$\mu_t - \frac{\tilde{\mu}_t^B}{\tilde{\beta}_t^B}\beta_t = r_t + \frac{\beta_t^2}{1+r_t} - \frac{(\mu_t - r_t)^2}{1+r_t} - \frac{\beta_t(\mu_t - r_t)\mathbf{E}_{t-1}\left[(\Delta Z_t)^3\right]}{1+r_t} .$$

Despite the more complicated form, we can find some analogy to the diffusion setup. The expected net return under the new measure is equal to the sum of the risk-free rate, the normalized variance, the normalized squared excess return and another term including a third moment of the basis martingale. Besides the scaling with $(1 + r_t)$ the first two summands correspond perfectly to the diffusion setup. In the corresponding continuous setting the drift of the asset is equal to the sum of the short rate of interest and the squared volatility, i.e. $r + \sigma^2$; see (3.20). However, there is no equivalent to the third summand. The fourth term might be interpreted as a correction term due to the higher moments which have to be taken into account in the discrete setup.

3.5.2 Binomial Tree

In this section we show explicitly how the binomial tree can be embedded in the discrete setup. Without loss of generality we consider a one-period model consisting of one risky asset and one riskless asset with two successor states. The risky asset S has the payoffs $S_0 u$ and $S_0 d$ at $t = 1$. The MMA with $B_0 = 1$ has the deterministic payoff $B_1 = (1 + r)$ at $t = 1$. The physical probability for the up-state is given by p, for the down-state by $1 - p$. Since there is only one risky asset with two possible states in $t = 1$, we need only one basis martingale to describe the dynamics of S. In our setting the basis martingale Z is given by

$$Z(\omega) = \Delta Z(\omega) = \begin{cases} \sqrt{\frac{1-p}{p}} & \text{if } \omega = \text{up}, \\ -\sqrt{\frac{p}{1-p}} & \text{if } \omega = \text{down}, \end{cases}$$

since we have assumed an expectation of zero and a variance of one for ΔZ. The price of the risky asset S_1 at $t = 1$ is given by

$$S_1 = S_0(1 + \mu + \beta \Delta Z), \tag{3.31}$$

where μ is the expected net return of S and β the standard deviation of S_1/S_0. To make things more explicit, we consider (3.31) for both states:

$$u = 1 + \mu + \beta\sqrt{\frac{1-p}{p}}, \tag{3.32}$$

$$d = 1 + \mu - \beta\sqrt{\frac{p}{1-p}}. \tag{3.33}$$

Subtracting (3.33) from (3.32) and rearranging terms leads to an expression for β:

$$\beta = \frac{u-d}{\sqrt{\frac{1-p}{p}} + \sqrt{\frac{p}{1-p}}}. \tag{3.34}$$

Plugging this into (3.32) yields

$$\mu = \frac{(u-1)\sqrt{\frac{p}{1-p}} + (d-1)\sqrt{\frac{1-p}{p}}}{\sqrt{\frac{1-p}{p}} + \sqrt{\frac{p}{1-p}}}. \tag{3.35}$$

The market price of risk λ is unique in this setup and simply given by the risk premium $\mu - r$ over one unit of risk (i.e. β). This follows from the drift restriction in (3.16). Plugging (3.34) and (3.35) into the drift restriction $\mu = r + \beta\lambda$, we can compute λ as

$$\lambda = \frac{(u-(1+r))\sqrt{\frac{p}{1-p}} + (d-(1+r))\sqrt{\frac{1-p}{p}}}{u-d}.$$

Thus, λ is constant in this simple setup.

Let us now consider the risk-neutral probabilities corresponding to the MMA as the numeraire: It can easily be derived that $\hat{p}$, the risk-neutral probability for the up-state, is given by

$$\hat{p} = \frac{1+r-d}{u-d}.$$

Hence, the change of measure from P to $\hat{P}$ in the up-state is simply

$$\frac{\hat{p}}{p} = \frac{1+r-d}{p(u-d)}$$

and in the down-state

$$\frac{1-\hat{p}}{1-p} = \frac{u-(1+r)}{(1-p)(u-d)}.$$

On the other hand, we know from 3.3.2 that the change of measure in the discrete setting is computed as

$$d\hat{P} = (1 - \lambda \Delta Z) dP. \tag{3.36}$$

The following calculations show that (3.36) is indeed true. We get

$$\begin{aligned} 1 - \lambda\, \Delta Z(\omega = \text{up}) &= 1 - \frac{(u-(1+r))\sqrt{\frac{p}{1-p}} + (d-(1+r))\sqrt{\frac{1-p}{p}}}{u-d}\sqrt{\frac{1-p}{p}} \\ &= \frac{u-d-(u-(1+r))-(d-(1+r))\frac{1-p}{p}}{(u-d)} \\ &= \frac{1+r-d}{p(u-d)} \\ &= \frac{\hat{p}}{p}. \end{aligned}$$

Analogously we find

$$1 - \lambda\, \Delta Z(\omega = \text{down}) = \frac{1-\hat{p}}{1-p},$$

which proves (3.36).

3.5.3 Two Correlated Assets

As the first example in continuous time, we consider a complete market consisting of two correlated risky basis assets and the MMA. We study the dynamics under the risk-neutral measure and the market prices of risk. The two basis assets follow the dynamics under the physical measure P given by

$$\begin{aligned} dS_t &= \mu_t S_t dt + \sigma_t S_t dW_t^{(S)}, \\ d\tilde{S}_t &= \tilde{\mu}_t \tilde{S}_t dt + \tilde{S}_t \tilde{\sigma}_t dW_t^{(\tilde{S})}, \end{aligned}$$

where $dW_t^{(S)} \bullet dW_t^{(\tilde{S})} = \gamma\, dt$. The parameter γ denotes an arbitrary, but constant correlation of the increments of the Brownian motion. A Cholesky decomposition into a two-dimensional Brownian motion with uncorrelated increments $W = (W^{(1)}, W^{(2)})$ yields

$$\begin{aligned} dS_t &= \mu_t S_t dt + \sigma_t \gamma S_t dW_t^{(2)} + \sigma_t \sqrt{1-\gamma^2} S_t dW_t^{(1)}, \\ d\tilde{S}_t &= \tilde{\mu}_t \tilde{S}_t dt + \tilde{S}_t \tilde{\sigma}_t dW_t^{(2)} \end{aligned}$$

(see, e.g. Shreve [40]). Using the MMA B as the numeraire provides the dynamics of the discounted processes

$$d(S_t/B_t) = (\mu_t - r_t)(S_t/B_t)dt + \sigma_t\gamma(S_t/B_t)dW_t^{(2)} + \sigma_t\sqrt{1-\gamma^2}(S_t/B_t)dW_t^{(1)}$$
$$d(\tilde{S}_t/B_t) = (\tilde{\mu}_t - r_t)(\tilde{S}_t/B_t)dt + (\tilde{S}_t/B_t)\tilde{\sigma}_t dW_t^{(2)}.$$

The market price of risk vector must be chosen such that the drift of both discounted processes vanishes after the change of measure. Thus, it consists of the two components

$$\lambda_t^{(1)} = \frac{(\mu_t - r_t)\tilde{\sigma}_t - \gamma\sigma_t(\tilde{\mu}_t - r_t)}{\sigma_t\tilde{\sigma}_t\sqrt{1-\gamma^2}},$$
$$\lambda_t^{(2)} = \frac{\tilde{\mu}_t - r_t}{\tilde{\sigma}_t}.$$

We now examine how these formulae fit into the intuitive idea of the market price of risk, being the excess return per unit of risk. The market prices of risk $\lambda^{(1)}$ and $\lambda^{(2)}$ correspond to the risk factors $W^{(1)}$ and $W^{(2)}$. We can equivalently look at the basis assets S and $\tilde{S}$ as risk factors and compute the market prices of risk for them. We expect the well-known form for the market price of risk of an asset, which is indeed the case: The market price of risk for asset $\tilde{S}$ is given by

$$\lambda_t^{(\tilde{S})} = \frac{\tilde{\mu}_t - r_t}{\tilde{\sigma}_t},$$

and the market price of risk for asset S – including both risk factors – is given by

$$\lambda_t^{(S)} = \gamma\lambda_t^{(2)} + \sqrt{1-\gamma^2}\lambda_t^{(1)} = \frac{\mu_t - r_t}{\sigma_t}.$$

This is exactly the desired result consistent with the economic intuition: The market price of risk is the excess return, i.e. the return of the asset minus the risk-free rate, over one unit of risk (given by the volatility). This result is in particular independent of the choice of representation. We get the same market prices of risk for the assets S and $\tilde{S}$, choosing the Cholesky decomposition such that the second asset is driven by two sources of risk and the first by one. When considering uncorrelated Brownian motions, we also find the same market prices of risk for the assets S and $\tilde{S}$.

3.5.4 Stochastic Volatility Setup

The most prominent examples of incomplete markets are SV models. We now derive the risk-neutral dynamics and the market prices of risk. The dynamics under the physical measure P of the asset S and the SV process σ are given by

$$\begin{aligned} dS_t &= \mu_t S_t dt + \sigma_t S_t dW_t^{(S)}, \\ d\sigma_t &= l(t,\sigma)dt + c(t,\sigma)dW_t^{(\sigma)}, \end{aligned}$$

where $dW_t^{(S)} \bullet dW_t^{(\sigma)} = \gamma\, dt$. The Cholesky decomposition into independent components of a two-dimensional Brownian motion $W = (W^{(1)}, W^{(2)})'$ leads to

$$\begin{aligned} dS_t &= \mu_t S_t dt + \sigma_t S_t dW_t^{(1)}, \\ d\sigma_t &= l(t,\sigma)dt + c(t,\sigma)\left(\gamma dW_t^{(1)} + \sqrt{1-\gamma^2} dW_t^{(2)}\right). \end{aligned}$$

The dynamics of the discounted asset price process are given by

$$d(S_t/B_t) = (\mu_t - r_t)(S_t/B_t)dt + \sigma_t (S_t/B_t) dW_t^{(1)}.$$

The market prices of risk for the risk factors $W^{(1)}$ and $W^{(2)}$ must be chosen such that the process S_t/B_t has zero drift under the new measure. This yields that $\lambda_t^{(1)} = (\mu_t - r_t)/\sigma_t$, $\lambda_t^{(2)}$ is arbitrary. The change of measure to $\hat{P}$ is given by

$$\begin{aligned} d\hat{W}_t^{(1)} &= \lambda_t^{(1)} dt + dW_t^{(1)} = \frac{\mu_t - r_t}{\sigma_t} dt + dW_t^{(1)}, \\ d\hat{W}_t^{(2)} &= \lambda_t^{(2)} dt + dW_t^{(2)}. \end{aligned}$$

The resulting dynamics under $\hat{P}$ are

$$\begin{aligned} dS_t &= r_t S_t dt + \sigma_t S_t d\hat{W}_t^{(1)}, \\ d\sigma_t &= l^*(t,\sigma)dt + c(t,\sigma)\left(\gamma d\hat{W}_t^{(1)} + \sqrt{1-\gamma^2} d\hat{W}_t^{(2)}\right), \end{aligned}$$

where the drift coefficient of the volatility process under $\hat{P}$ is

$$l^*(t,\sigma) = l(t,\sigma) - c(t,\sigma)\left(\gamma\lambda_t^{(1)} + \sqrt{1-\gamma^2}\lambda_t^{(2)}\right).$$

As before, we can switch from the risk factors $W^{(1)}$ and $W^{(2)}$ to the asset S and the state variable σ and calculate the market prices of risk for them. The market price of risk for the asset S is equal to

$$\lambda_t^{(S)} = \frac{\mu_t - r_t}{\sigma_t} \equiv \lambda_t^{(1)}$$

which is no surprise. Analogously, we can derive the market price of risk for the volatility – including both sources of risk – as

$$\lambda_t^{(\sigma)} = \gamma\lambda_t^{(1)} + \sqrt{1-\gamma^2}\lambda_t^{(2)} = \gamma\,\frac{\mu_t - r_t}{\sigma_t} + \sqrt{1-\gamma^2}\lambda_t^{(2)}.$$

Since the market is incomplete, there remains one degree of freedom in the choice of $\lambda^{(\sigma)}$, as $\lambda^{(2)}$ is arbitrary. This is for example further investigated in an equilibrium framework in Pham and Touzi [35].

3.6 Conclusion

In this chapter we have investigated the similarities and differences between the continuous and discrete setup. In the discrete setup basically the same properties with economic meaning such as drift restriction, market prices of risk and the relation to the risk-neutral measures hold as in the diffusion setup. However, we have seen that in some sense diffusion models are easier to handle, especially when considering normalized price processes. There is even a compact form for the drift under every EMM: It is equal to the sum of the risk-free rate of interest and the covariation of the asset return and the numeraire returns. For every numeraire the change of measure induced is given explicitly by the drift restrictions, also involving the market prices of risk. These features are missing in the discrete setting. Furthermore, the following properties of a diffusion model do not carry over to the discrete setup: First, the equivalent measure obtained by an arbitrary predictable process α is always a probability measure which can fail in a discrete model. Second, the change of measure in the diffusion setup only involves the drift of the stochastic process, but not the covariance. This does not hold in the discrete model, since in contrast to the new Brownian motion the new basis martingales are no longer conditionally uncorrelated. Thus, we have depicted some features of the diffusion model which do not necessarily hold in a generic discrete setting.

4

Valuation of Power Options

4.1 Introduction

This chapter deals with the pricing of certain types of exotic options, called "power options" and "powered options". The special feature of these contracts is that, compared with plain vanilla options, in the first case the stock price in the payoff function is replaced by the stock price raised to some power, and in the latter case the option payoff is raised to some power. These contracts generalize the special case of a piecewise-linear payoff for plain vanilla contracts. Without loss of generality we only deal with calls, since puts can be priced similarly. If the exponent of a power call option is greater (smaller) than one, the payoff and consequently the price of such contracts is greater (smaller) than the corresponding plain vanilla call option.

To illustrate the typical contractual features, we consider for instance the case of a power call option with exponent greater than one. Compared with a standard call option, the power option has a payoff that provides a greater leverage effect caused by the potential for a higher payoff at maturity. On the other hand, such a contract has a higher initial premium than the corresponding standard option. In practice such contracts are usually capped at a certain level to bound the risk for the short party of a power option. We discuss the pricing equation for such a capped contract as well.

The valuation of power options is based on suitable choices of artificial measures as is the standard approach for European options. In general, however, due to the more complex payoff structure the identification with associated numeraires fails. The method, which enables us to derive a concise pricing equation nevertheless, has already been discussed in Chap. 2 and will be applied here. The two settings we consider are the BS model and an SV model.

There is a wide range of literature dealing with power options in the BS setting. Among others there could be named the works by Tompkins [41] and Furlan and Pechtl [26] which address hedging strategies as well. In contrast to them, we focus on the change of measure in the BS setup. We know from 2.5.1 that in the BS setup there exists a numeraire portfolio corresponding to the change of measure since the market is complete, but it is not obvious what it looks like. We will compute this portfolio in detail, thereby gaining some economic insight.

Further, we deal with SV models which have become a class of widely accepted models for pricing in the last two decades. Some of the most prominent models were suggested by Heston [28], Schöbel and Zhu [37], and Bakshi, Cao, and Chen [2], but these papers mainly deal with plain vanilla options. There is not much literature on exotic options in SV models; Zhu [43] analyzes the pricing of several exotic contracts under SV and stochastic interest rates; Bakshi and Madan [3] come across a squared power payoff in a general diffusion setup. The application of the change of measure technique to power options in SV models is of general interest, as it further extends the class of exotic claims that allow for a closed-form solution in an SV setup. This technique also works for other types of exotic options, such as product and quotient options which are briefly discussed in Zhu [43]. The key feature of this approach is that we are able to derive a quasi-closed form pricing equation for three types of power options in SV models which have not been reported in the existing literature so far.

This chapter is based on a paper by Esser [20] on general valuation principles of arbitrary payoffs and applications to power options.

4.2 General Pricing Equation

4.2.1 Power Option

The payoff at time T of a power option with strike K on an underlying S is denoted by Pow_T and given by

$$\mathrm{Pow}_T = (S_T^a - K)^+, \quad a \in \mathbb{R}^+,$$

where $z^+ := \max[z, 0]$. The price of the power option at time t is calculated using risk-neutral valuation:

$$\begin{aligned}\text{Pow}_t &= \frac{B_t}{B_T}\hat{\mathbf{E}}_t\left[(S_T^a - K)^+\right] \\ &= \frac{B_t}{B_T}\hat{\mathbf{E}}_t\left[S_T^a I(S_T^a > K)\right] - K\frac{B_t}{B_T}\hat{\mathbf{E}}_t\left[I(S_T^a > K)\right].\end{aligned} \tag{4.1}$$

The valuation of the second summand does not cause a problem since strike and interest rate are constants, which means that we only have to compute the $\hat{P}$-probability of the corresponding event. The more challenging issue raised in (4.1) is how to compute the expectation $\hat{\mathbf{E}}_t\left[S_T^a I(S_T^a > K)\right]$.

For $a = 1$, i.e. in the case of a plain vanilla call, we follow the standard approach discussed in 2.3.3. That is, we choose the underlying asset as the numeraire in the first summand, thereby changing the measure from $\hat{P}$ to $\tilde{P}$. The measure $\tilde{P}$ is the EMM under which price processes normalized with the underlying asset S are martingales. The change of measure is given by (2.10). The problem then again reduces to calculating the expectation under $\tilde{P}$ of an indicator variable which is exactly the $\tilde{P}$-probability of the corresponding event shown in (2.11). For $a \neq 1$ one might be tempted to proceed analogously and switch from the MMA as the numeraire to S^a. However, S_t^a cannot be taken as a numeraire since it is not the price process of a traded asset for $a \neq 1$. This follows from the fact that the discounted asset price S_t/B_t and the discounted S_t^a (i.e. S_t^a/B_t) cannot be $\hat{P}$-martingales simultaneously. We can see this easily using Jensen's inequality as S^a is a strictly convex(concave) function for $a > 1 (a < 1)$.

Since the change of numeraire does not work here, we have to apply the general change of measure discussed in Sect. 2.4. We define the measure $P^{(a)}$ by (2.15), setting $g(S) = S^a$ and $N \equiv B$, i.e.

$$dP^{(a)} = \frac{S_T^a}{\hat{\mathbf{E}}_0\left[S_T^a\right]}\, d\hat{P} \quad \text{on } \mathcal{F}_T. \tag{4.2}$$

Hence, using (4.2) we can rewrite the first summand in (4.1) as

$$\begin{aligned}\frac{B_t}{B_T}\hat{\mathbf{E}}_t\left[S_T^a I(S_T^a > K)\right] &= \frac{B_t}{B_T}\hat{\mathbf{E}}_t\left[S_T^a\right]\hat{\mathbf{E}}_t\left[\frac{S_T^a}{\hat{\mathbf{E}}_t[S_T^a]} I(S_T^a > K)\right] \\ &= \frac{B_t}{B_T}\hat{\mathbf{E}}_t[S_T^a] P_t^{(a)}(S_T^a > K).\end{aligned} \tag{4.3}$$

Equation (4.3) is nothing but a special case of (2.16). So, the general pricing formula for the power option is given by

$$\mathrm{Pow}_t = \frac{B_t}{B_T}\hat{\mathbf{E}}_t[S_T^a]P_t^{(a)}(S_T^a > K) - K\frac{B_t}{B_T}\hat{P}_t(S_T^a > K). \qquad (4.4)$$

The first summand equals the price at t of the power of the stock at T times the artificial probability of exercise. Note that the price of the power of the stock at maturity (i.e. S_T^a) is not equal to S_t^a. This is due to the fact that S_t^a is not a price process of a traded asset for $a \neq 1$, recalling the argument following Jensen's inequality.

To summarize the key insight in the derivation above, we see that the change of measure in (4.2) works for every $a \in \mathbb{R}^+$. Nevertheless, the difference between the approach for plain vanilla options and power options becomes visible when interpreting the change of measure as a special choice of numeraire: For $a = 1$ the new measure can be interpreted as an equivalent martingale measure for the normalized price processes with S as the numeraire. For $a \neq 1$ the new measure $P^{(a)}$ computed in (4.2) does not correspond to S_t^a as the numeraire, since S_t^a is not a price process.

We can now apply the general results developed in Chap. 2 to interpret the measure $P^{(a)}$, distinguishing between complete and incomplete markets.

We have seen in 2.5.1 that in a complete market every measure, equivalent to the physical measure, is an EMM with respect to the portfolio replicating the Radon–Nikodym density. Setting $V_T = S_T^a$ in our model, we can rewrite the price of the payoff $F_T := S_T^a I(S_T^a > K)$ at t as

$$\begin{aligned} F_t &= \frac{B_t}{B_T}\hat{\mathbf{E}}_t[S_T^a I(S_T^a > K)] \\ &= V_t \mathbf{E}_t^{(a)}\left[\frac{S_T^a I(S_T^a > K)}{V_T}\right] \\ &= V_t P_t^{(a)}(S_T^a > K). \end{aligned} \qquad (4.5)$$

This is exactly the simplified pricing equation (4.3), standing in analogy to (2.20). Since $V_t = (B_t/B_T)\hat{\mathbf{E}}_t[V_T]$, the change of measure, given in (4.2), can equivalently be represented by the ratio of the numeraires V and B:

$$\frac{dP^a}{d\hat{P}} = \frac{S_T^a}{\hat{\mathbf{E}}_0[S_T^a]} = \frac{V_T/V_0}{B_T/B_0}.$$

Thus, $P^{(a)}$ is an EMM with respect to the numeraire V.

We will discuss the measure $P^{(a)}$ for the power option in the BS setting explicitly in 4.3.1. In this case it is possible to determine the associated numeraire portfolio V explicitly.

Let us now consider an incomplete market. For a fixed risk-neutral measure $\hat{P}$ the general scenario has been discussed in 2.5.2. If and only if $S_T^{(a)}$ is attainable, the measure $P^{(a)}$ is an EMM, namely corresponding to V, where V denotes the replicating portfolio $V_T = S_T^a$. However, the payoff S_T^a will in general not be attainable in an incomplete market, which implies that there does not exist an attainable numeraire corresponding to $P^{(a)}$. This in turn means that $P^{(a)}$ is in general not an EMM in an incomplete setup. The issue attainability will be discussed further in the SV setup in 4.3.2.1.

4.2.2 Powered Option

The payoff at time T of a powered option with strike K on an underlying asset S, denoted by Powd_T, is given by

$$\text{Powd}_T = \max[(S_T - K), 0]^a = (S_T - K)^a I(S_T > K), \quad a \in \mathbf{N}.$$

The payoff can be rewritten as

$$\text{Powd}_T = \sum_{j=0}^{a} \binom{a}{j} S_T^{a-j} (-K)^j I(S_T > K).$$

The price at time t can be calculated using risk-neutral valuation and appropriate changes of measure:

$$\begin{aligned}
\text{Powd}_t &= \frac{B_t}{B_T} \hat{\mathbf{E}}_t \left[\max[(S_T - K), 0]^a\right] \\
&= \frac{B_t}{B_T} \hat{\mathbf{E}}_t \left[\sum_{j=0}^{a} \binom{a}{j} S_T^{a-j} (-K)^j I(S_T > K)\right] \\
&= \frac{B_t}{B_T} \sum_{j=0}^{a} \binom{a}{j} (-K)^j \hat{\mathbf{E}}_t \left[S_T^{a-j} I(S_T > K)\right] \\
&= \frac{B_t}{B_T} \sum_{j=0}^{a} \binom{a}{j} (-K)^j \hat{\mathbf{E}}_t [S_T^{a-j}] P_t^{(a-j)} (S_T > K), \qquad (4.6)
\end{aligned}$$

where in analogy to the derivation in 4.2.1 we set

$$dP^{(a-j)} = \frac{S_T^{a-j}}{\hat{\mathbf{E}}_0 \left[S_T^{a-j}\right]} d\hat{P} \text{ on } \mathcal{F}_T.$$

Note that in certain textbooks (e.g. Zhang [42]) the following different payoff structure is discussed:

$$\mathrm{Spow}_T = \max[(S_T - K)^a, 0], \quad a \in \mathbb{N}.$$

This is called the symmetric power option in contrast to the asymmetric payoff structure described before. The symmetric power option payoff coincides with the above defined powered option for odd exponents. For even exponents $a = 2L, L \in \mathbb{N}$, this contract has a symmetric payoff

$$\mathrm{Spow}_T = (S_T - K)^{2L} = \sum_{j=0}^{2L} \binom{2L}{j} S_T^{2L-j} (-K)^j.$$

Its price at t is then simply given by

$$\mathrm{Spow}_t = \frac{B_t}{B_T} \sum_{j=0}^{2L} \binom{2L}{j} (-K)^j \hat{\mathbf{E}}_t[S_T^{2L-j}].$$

A formula for $a \in \mathbb{R}^+$ in the BS model can be found in Furlan/Pechtl [26].

4.2.3 Capped Power Option

In practice the payoff of a power option is typically capped to limit the risk for the seller of the option. Therefore, we introduce an upper bound $\bar{C}$. If the payoff of the power option without cap exceeds this upper bound, it is set equal to $\bar{C}$, i.e. the payoff structure is given by the following equation:

$$\begin{aligned}
\mathrm{Powc}_T &= \min\left[(S_T^a - K)^+; \bar{C}\right] \\
&= \min\left[(S_T^a - K) I(S_T^a > K); \bar{C}\right] \\
&= (S_T^a - K) I(\bar{C} + K > S_T^a > K) + \bar{C} I(S_T^a \geq \bar{C} + K) \\
&= (S_T^a - K) I(S_T^a > K) - (S_T^a - (\bar{C} + K)) I(S_T^a \geq \bar{C} + K).
\end{aligned}$$

This payoff structure can be interpreted as a portfolio consisting of a power option long with strike K and a power option short with strike $\bar{C} + K$ with the same exponent a and the same maturity T. The pricing equation is then given by

$$\begin{aligned}
\mathrm{Powc}_t = &\frac{B_t}{B_T} \hat{\mathbf{E}}_t[S_T^a] \left[P_t^{(a)}(S_T^a > K) - P_t^{(a)}\left(S_T^a \geq \bar{C} + K\right)\right] \\
&- \frac{B_t}{B_T} \left[K \hat{P}_t(S_T^a > K) - (\bar{C} + K) \hat{P}_t\left(S_T^a \geq \bar{C} + K\right)\right]. \quad (4.7)
\end{aligned}$$

4.3 Examples

4.3.1 Black–Scholes Model

We assume that the underlying asset S follows a geometric Brownian motion with drift r and volatility σ under the risk-neutral measure $\hat{P}$. Then, $X_T = \ln S_T$, given X_t, is normally distributed with expectation $\ln S_t + (r - \sigma^2/2)(T - t)$ and variance $\sigma^2(T - t)$.

4.3.1.1 Pricing Equation

Now we want to calculate the price Pow_t explicitly. The second summand in (4.4) follows directly from BS by substituting $(\ln K)/a$ for $\ln K$, i.e.

$$\begin{aligned} -Ke^{-r(T-t)}\hat{P}_t(S_T^a > K) &= -Ke^{-r(T-t)}\hat{P}_t(\ln S_T > (\ln K)/a) \\ &= -Ke^{-r(T-t)}\mathcal{N}(h_2), \end{aligned}$$

where $\mathcal{N}$ denotes the distribution function of the standard normal distribution and

$$h_2 = \frac{\ln\left(S_t/(K^{1/a})\right) + (r - \frac{1}{2}\sigma^2)(T-t)}{\sigma\sqrt{T-t}} .$$

The coefficient of the first summand in (4.4) is computed using the formula for the expectation of a log-normally distributed random variable $\bar{X}$, that is calculated as $\mathbf{E}[\exp(\bar{X})] = \exp(\mathbf{E}\bar{X} + \frac{1}{2}\text{Var}\bar{X})$. Since S_T^a is log-normally distributed, given S_t, it holds that

$$e^{-r(T-t)}\hat{\mathbf{E}}_t[S_T^a] = e^{-r(T-t)}S_t^a \exp\left(a\left(r - \frac{1}{2}\sigma^2\right)(T-t) + \frac{1}{2}a^2\sigma^2(T-t)\right).$$

The artificial probability $P^{(a)}$ is defined by the Radon–Nikodym derivative with respect to the risk-neutral measure $\hat{P}$:

$$\frac{S_T^a}{\hat{\mathbf{E}}_0[S_T^a]} = \frac{S_0^a \exp\left(a(r - \frac{1}{2}\sigma^2)T + a\sigma\hat{W}_T\right)}{S_0^a \exp\left(a(r - \frac{1}{2}\sigma^2)T + \frac{1}{2}a^2\sigma^2 T\right)} = \exp\left(-\frac{1}{2}(a\sigma)^2 T + a\sigma\hat{W}_T\right).$$

According to Girsanov's theorem, the Radon–Nikodym derivative for a one-dimensional diffusion process is of the form

$$\exp\left(-\int_0^T \alpha_t d\hat{W}_t - \frac{1}{2}\int_0^T \alpha_t^2 dt\right)$$

for a predictable process α; see 3.3.1, Equation (3.5). In our scenario, this yields

$$\alpha_t = -a\sigma \quad \forall t.$$

Since the change of measure relates the Brownian motions $W_t^{(a)}$ and $\hat{W}_t$ by

$$dW_t^{(a)} = d\hat{W}_t - a\sigma dt,$$

the drift of the underlying asset under $P^{(a)}$ is increased by $a\sigma^2$. Thus, the dynamics of S and $\ln S$ under $P^{(a)}$ are

$$\begin{aligned} dS_t/S_t &= (r + a\sigma^2)dt + \sigma dW_t^{(a)} \\ d\ln S_t &= (r + (a - \frac{1}{2})\sigma^2)dt + \sigma dW_t^{(a)}, \end{aligned} \tag{4.8}$$

respectively. Thus, $\ln S_T$, given $\ln S_t$, is normally distributed with expectation

$$\ln S_t + \left(r + \left(a - \frac{1}{2}\right)\sigma^2\right)(T - t)$$

and variance $\sigma^2(T-t)$ under $P_t^{(a)}$. This leads to the following formula for the first artificial probability in (4.4):

$$P_t^{(a)}(S_T^a > K) = P_t^{(a)}(\ln S_T > (\ln K)/a) = \mathcal{N}(h_1),$$

where

$$h_1 = \frac{\ln(S_t/K^{1/a}) + (r + (a - \frac{1}{2})\sigma^2)(T-t)}{\sigma\sqrt{T-t}} = h_2 + a\sigma\sqrt{T-t}.$$

Hence, the following pricing formula for a power option in the BS setting holds:

$$\begin{aligned} \text{Pow}_t &= e^{-r(T-t)}\hat{\mathbf{E}}_t\left[S_T^a\right] P_t^{(a)}\left(S_T^a > K\right) - Ke^{-rT}\hat{P}_t(S_T^a > K) \\ &= S_t^a e^{(a-1)(r+\frac{1}{2}a\sigma^2)(T-t)}\mathcal{N}(h_1) - Ke^{-r(T-t)}\mathcal{N}(h_2). \end{aligned} \tag{4.9}$$

As we can see from (4.9), for $a = 1$ we end up with the classical BS fomula for a plain vanilla call option.

Analogously, the price of the powered option is given by

$$\begin{aligned} \text{Powd}_t &= e^{-r(T-t)}\sum_{j=0}^{a}\binom{a}{j}(-K)^j\hat{\mathbf{E}}_t\left[S_T^{a-j}\right] P_t^{(a-j)}\left(S_T > K\right) \\ &= \sum_{j=0}^{a}\binom{a}{j} S_t^{a-j}(-K)^j e^{(a-j-1)(r+\frac{1}{2}(a-j)\sigma^2)(T-t)}\mathcal{N}(d^{(a-j)}), \end{aligned}$$

where

$$d^{(a-j)} = \frac{\ln(S_t/K) + (r + (a - j - \frac{1}{2})\sigma^2)(T - t)}{\sigma\sqrt{T - t}}.$$

Again for $a = 1$ we are back in the BS model for a standard call.

The price of a capped power option is

$$\begin{aligned}\mathrm{Powc}_t &= S_t^a e^{(a-1)(r+\frac{1}{2}a\sigma^2)(T-t)} \left(\mathcal{N}(h_1) - \mathcal{N}(\bar{h}_1)\right) \\ &\quad - e^{-r(T-t)}\left(K\mathcal{N}(h_2) - (\bar{C} + K)\mathcal{N}(\bar{h}_2)\right),\end{aligned}$$

where $\bar{h}$ is defined by substituting K for $\bar{C} + K$ in the formulae for h, i.e.

$$\bar{h}_1 = \frac{\ln(S_t/(\bar{C} + K)^{1/a}) + (r + (a - \frac{1}{2})\sigma^2)(T - t)}{\sigma\sqrt{T - t}} = \bar{h}_2 + a\sigma\sqrt{T - t}.$$

4.3.1.2 Numeraire Portfolio

After having shown the derivation of the pricing equation, we now examine the measure $P^{(a)}$ in more detail. Since the market is complete in the BS setup, there exists a traded asset that serves as the numeraire V for the measure $P^{(a)}$, i.e. the price processes normalized by V are $P^{(a)}$-martingales. Hence, choosing a positive self-financing portfolio V such that $V_T = S_T^a$, equation (4.3) is equivalent to (4.5). Now we want to compute the numeraire portfolio V associated with the artificial measure $P^{(a)}$. The price of the portfolio V can be found by calculating the delta of the claim with payoff $S_T^a = V_T$, since delta represents the number of units of the underlying asset in the hedge portfolio with the remaining funds going into the MMA. The price of the portfolio V at time t is given by

$$V_t = e^{-r(T-t)}\hat{\mathbf{E}}_t[S_T^a] = e^{-r(T-t)} S_t^a \exp\left(a\left(r - \frac{1}{2}\sigma^2\right)(T - t) + \frac{1}{2}a^2\sigma^2(T - t)\right),$$

and the delta is equal to

$$\Delta_t \equiv \frac{\partial V_t}{\partial S_t} = a e^{-r(T-t)} S_t^{a-1} \exp\left(a\left(r - \frac{1}{2}\sigma^2\right)(T - t) + \frac{1}{2}a^2\sigma^2(T - t)\right) = \frac{a}{S_t} V_t.$$

Thus, the total amount invested in the underlying asset to replicate the claim is given by aV_t, and $(1 - a)V_t$ is the total amount invested in the MMA. This means that at any point in time t a constant proportion a of the claim's value is invested in the stock. To put it differently, the relative portfolio weight $\Delta_t S_t / V_t$ of the underlying asset in the hedge portfolio is simply given by the constant exponent a. This means money has to be borrowed for an exponent greater than one. The price of the hedge portfolio is then equal to

$$V_t = \Delta_t S_t + \phi_t B_t = e^{-r(T-t)} \hat{\mathbf{E}}_t[S_T^a], \qquad \text{where} \qquad \phi_t = \frac{(1-a)V_t}{B_t}.$$

This is the numeraire portfolio V, associated with the measure $P^{(a)}$ in the BS setup.

Taking now the portfolio $V_t = \Delta_t S_t + \phi_t B_t$ as the numeraire corresponding to the EMM $P^{(a)}$, it must be possible to derive the dynamics under $P^{(a)}$ by considering the stock price process normalized with the portfolio V. This has been discussed in detail in 3.4.3.1. With $n = m = 1$ and $x^1 S^1/V = a$, we get the same drift coefficient $r + a\sigma^2$ from (3.29) as derived here in (4.8). Thus, we have seen in this example of a complete market how the change of measure, using a Radon–Nikodym derivative, can be rewritten in terms of a numeraire portfolio. This in turn corresponds to the change of measure making price processes normalized with this numeraire portfolio into martingales.

4.3.2 Stochastic Volatility Models

4.3.2.1 Attainable Payoffs

Stochastic volatility models are prominent examples for incomplete markets. In the context of incomplete markets the issue of attainability is of major interest for pricing and hedging purposes. We now restrict ourselves to claims in the SV model where the payoffs are functions of the terminal stock only. The question regarding for which claims a replicating portfolio exists can be answered as follows: All claims that lie in the subspace formed by the self-financing trading strategies using the risky basis assets and the MMA are attainable. However, the problem remains for a given payoff to find out whether it is attainable or not. For a detailed discussion of this subject see Branger, Esser, and Schlag [9]. There, the following statement is proved:

In the SV setup only a claim which is affine-linear in the underlying asset is attainable using a self-financing portfolio strategy.

The key idea of the proof is briefly summarized in the following.

It can be observed by looking at the dynamics of an attainable claim in the SV setup that the claim must be independent of the additional source of risk at every point in time. Thus, the claim is a function of time and underlying stock only. This makes sound sense, because volatility is not traded and we therefore cannot compensate the non-traded volatility risk in a hedge by using only the underlying asset and the MMA as hedge instruments.

When considering the partial differential equation in the SV model which is satisfied by the claim, we find that it simplifies considerably, since the claim is independent of volatility. Moreover, we can conclude from the partial differential equation that the contract must be linear in S and t.

Thinking of the variety of possible trading strategies, the result is quite surprising since one would have expected a wider class of attainable payoffs at first sight. It is worth noting that an analogous result is not true in the discrete setup. A simple example of a non-linear claim which can be replicated in a discrete tree model with SV is provided in Branger, Esser, and Schlag [9].

Hence, we can conclude that S_T^a for $a \neq 1$ is not attainable in the SV model. This in turn means that there is no numeraire portfolio corresponding to the artificial measure $P^{(a)}$ in the SV setup.

4.3.2.2 Quasi-Closed Form Pricing Equation

The basic pricing approach discussed in Sect. 4.2 also carries over to more general option pricing models. In the case of SV models we are able to derive a quasi-closed form pricing equation using the technique presented in Sect. 4.2. The only thing left to do is to calculate the artificial probabilities using Fourier transform techniques as in Schöbel and Zhu [37].

Consider the dynamics of the underlying asset S and the instantaneous volatility σ under a risk-neutral measure $\hat{P}$:

$$\begin{aligned} dS_t &= r_t S_t dt + \sigma_t S_t d\hat{W}_t^S, \\ d\sigma_t &= l^*(t,\sigma)dt + c(t,\sigma)d\hat{W}_t^\sigma, \end{aligned}$$

where $d\hat{W}_t^S \bullet d\hat{W}_t^\sigma = \gamma dt$. Note that $\hat{P}$ is no longer unique because the market is incomplete since volatility is not a traded asset.

Popular choices for the volatility process are the Ornstein–Uhlenbeck process (as in Schöbel and Zhu [37]), the square-root process for the instantaneous variance (as in Heston [28]), or similar dynamics as discussed in Zhu [43]. A further possible extension would be to include a stochastic process for the short-term interest rate, leading to a very flexible model of the type studied by Bakshi, Cao, and Chen [2].

The pricing equation for the power option is given by (4.4). In contrast to BS we now have to calculate the artificial probabilities via a Fourier inversion approach.

Let $K^* = (\ln K)/a$. We know from Shephard [39] that the following formulae hold:

$$\hat{P}_t(S_T^a > K) = \hat{P}_t(\ln S_T > K^*) = \frac{1}{2} + \frac{1}{\pi}\int_0^\infty \Re\left(\frac{\exp(-ikK^*)\hat{\chi}_{\ln S_T,t}(k)}{ik}\right) dk,$$

$$P_t^{(a)}(S_T^a > K) = P_t^{(a)}(\ln S_T > K^*) = \frac{1}{2} + \frac{1}{\pi}\int_0^\infty \Re\left(\frac{\exp(-ikK^*)\chi_{\ln S_T,t}^{(a)}(k)}{ik}\right) dk,$$

where $\hat{\chi}, \chi^{(a)}$ denote the Fourier transforms of $\hat{P}, P^{(a)}$, respectively, and $\Re$ stands for the real part of a complex number. The Fourier transforms are defined by

$$\begin{aligned}
\hat{\chi}_{\ln S_T,t}(k) &= \hat{\mathbf{E}}_t\left[e^{ik\ln S_T}\right] = \hat{\mathbf{E}}_t\left[e^{ikX_T}\right], \\
\chi_{\ln S_T,t}^{(a)}(k) &= \mathbf{E}_t^{(a)}\left[e^{ik\ln S_T}\right] \\
&= \hat{\mathbf{E}}_t\left[\frac{e^{aX_T}}{\hat{\mathbf{E}}_t[e^{aX_T}]}e^{ikX_T}\right] \\
&= \frac{1}{\hat{\mathbf{E}}_t[S_T^a]}\hat{\mathbf{E}}_t\left[e^{(a+ik)X_T}\right],
\end{aligned} \tag{4.10}$$

where $X_T = \ln S_T$. A detailed simplification of the expressions above can be found in the Appendix, Sect. A.1. So far, the derivation holds for any SV model. To derive an explicit solution for the conditional expectations we have to specify the volatility process. We restrict ourselves to the case of an Ornstein–Uhlenbeck process for the stochastic volatility. The characteristic functions can be represented in a closed form using the Feynman–Kac theorem; see the Appendix, Sect. A.2.

The only problem that remains is to calculate the conditional expectation of S_T^a in formula (4.10). The computation of this expectation is straightforward, since

$$\hat{\mathbf{E}}_t[S_T^a] = \hat{\mathbf{E}}_t[\exp(ik\ln S_T)|_{k=-ia}] = \hat{\chi}_{\ln S_T,t}(-ia),$$

so that we can directly apply our results for the characteristic function $\hat{\chi}$ derived above.

The pricing equation for the powered option for $a \in \mathbf{N}$ is given by (4.6), where

$$P_t^{(a-j)}(\ln S_T > \ln K) = \frac{1}{2} + \frac{1}{\pi} \int_0^\infty \Re \left(\frac{\exp(-ikK)\chi_{\ln S_T,t}^{(a-j)}(k)}{ik} \right) dk, \; j = 0, \ldots, a,$$

in this setup. Again, $\chi^{(a-j)}$ denotes the Fourier transform of $P^{(a-j)}$, which can be calculated in the same way as shown above, replacing a by $(a-j)$ in (4.10), i.e.

$$\chi_{\ln S_T,t}^{(a-j)}(k) = \mathbf{E}_t^{(a-j)} \left[e^{ik \ln S_T} \right] = \frac{1}{\hat{\mathbf{E}}_t[S_T^{a-j}]} \hat{\mathbf{E}}_t[e^{(a-j+ik)X_T}].$$

All other steps of the derivation remain the same.

To calculate the price of the capped power option, given in (4.7), we additionally need the probabilities $\hat{P}_t(S_T^a \geq \bar{C} + K)$ and $P_t^{(a)}(S_T^a \geq \bar{C} + K)$, which are computed analogously substituting $K^* = \ln K/a$ by $C^* = \ln(\bar{C} + K)/a$.

4.4 Conclusion

In this chapter we have derived a general pricing equation for options, where either the payoff depends on some power of the stock price, or the payoff itself is raised to some power. After having developed the pricing formulae for three types of power options based on the general theory in Chap. 2, two applications have been given. They show the representation of the pricing equation in terms of artificial probabilities for three types of power options in the BS setup and for SV models. In the BS setup the EMM and the corresponding numeraire portfolio are calculated explicitly.

Furthermore, quasi-closed form solutions for the power options in the SV setup are computed. This is well worth the effort, because it simplifies numerical applications considerably. The main advantage of these pricing equations is that they allow for a much faster valuation of power contracts than time-consuming simulation methods.

5

Modeling Feedback Effects Using Stochastic Liquidity

5.1 Introduction

This chapter deals with the modeling of asset liquidity. One aspect of liquidity includes the price impact involved in acquiring or liquidating a position. Our objective is to study the interaction between the trading strategy of a large investor, the asset price process, and liquidity in one single setup. There is a growing number of theoretical papers investigating the interaction of liquidity and trading strategies of large investors. Part of this literature considers optimal liquidation strategies for large portfolios, for instance Dubil [17], and Almgren and Chriss [1]. Recently, research has focused more and more on the modeling and hedging aspects that are introduced by liquidity and the presence of large traders. Cvitanić and Ma [14], Frey [22], Frey and Patie [23], and Liu and Yong [31] consider liquidity as an exogenously given source of risk. Frey and Stremme [25], Kampovsky and Trautmann [29], Papanicolaou and Sircar [34], and Schönbucher and Wilmott [38] serve as prominent examples taking into account equilibrium setups.

Our approach (see Esser and Mönch [21]) is a generalization of both the model of Frey [22] – where liquidity is constant – and the extension by Frey and Patie [23] – where liquidity is a deterministic function of the stock price. Modeling liquidity as a stochastic factor first of all incorporates random changes in market depth. Furthermore, we are able to model trading strategies for the large investor that do not depend exclusively on the stock price, but also on liquidity. Thus, the stochastic liquidity factor has two main effects: Firstly, it influences the trading strategy of the institutional investor and secondly, it has an impact on the degree to which the stock price reacts to the trading activity of the large investor.

Why is it interesting to treat liquidity as an autonomous source of risk? Consider for example an insurance company. If a natural disaster occurs, the insurance company may have to liquidate substantial amounts of assets in a short time, in order to compensate clients for their losses. Therefore, the risk management in an insurance company will allow the fund managers to invest a significant share of the portfolio only in highly liquid assets so that the company is able to meet possible obligations in time. Thus, in this example, portfolio managers have to take into account the liquidity risk associated with their investments, and have to liquidate their positions if the illiquidity of the assets under consideration exceeds a certain threshold.

In this chapter we analyze the behavior of stock prices under feedback effects theoretically and discuss two examples in the SL model. First, we briefly summarize the main ideas of the constant liquidity (CL) model. Then, the effective price process of the extended model including stochastic liquidity is derived, which takes feedback effects of trading strategies and of liquidity on the stock price dynamics into account. We end up with a general diffusion model including SV. The dynamics under a risk-neutral measure are calculated following the general theory of Chap. 3. Second, two examples of our new model are presented to show the additional features compared with the CL model. We introduce two feedback strategies and compare simulation paths for the BS and the SL model. The results underscore the importance of an SL factor. Furthermore, we illustrate the applicability of our framework by proposing a liquidity derivative. The claim under consideration is an insurance against the discount due to illiquidity when the large trader has to unwind a significant position in a single trade using a stop loss strategy.

5.2 The Liquidity Framework

5.2.1 Constant Liquidity

We begin with a brief summary of the paper of Frey [22]. He proposes a model setup which extends the BS asset price dynamics, introducing a constant liquidity parameter ρ. There exists a risky asset S (the stock) and a risk-free investment earning a zero interest rate (the MMA). There are no liquidity effects on the MMA; only the underlying asset S is affected by this source of risk. Further, there is a single large investor whose trading activity influences the price process of the underlying asset. The underlying asset follows the SDE

$$dS_t = \sigma S_{t-} dW_t^S + \rho S_{t-} d\phi_t^{(+)}.$$

S_{t-} denotes the left limit of S at t. This is relevant, since S is not necessarily continuous as $\phi^{(+)}$ is not necessarily continuous. $\phi^{(+)}$ denotes the right-continuous version of ϕ, where ϕ represents the number of stocks held by the large trader. The quantity $\rho \geq 0$ is a constant liquidity parameter. The term "liquidity parameter" which has been established in Frey's paper [22], is actually a misnomer: The parameter ρ measures illiquidity, since an increase in ρ means that liquidity is declining in the market. For $\rho = 0$ the model represents the standard BS setup with zero drift. $1/(\rho S)$ is called market depth, which is defined in this setup as "the order size that moves the price by one unit". Selling assets, i.e. $d\phi^{(+)} < 0$, causes a decrease in stock prices: The larger the parameter ρ, the higher is the effect on dS. In order to assure that the stock price process is staying positive, we have to assume that the jump size is bounded such that $\rho \Delta\phi^{(+)} > -1$.

The impact of the trading strategy on the price process is discussed for the case of a smooth strategy $\phi \equiv \phi(t, S)$, $\phi \in \mathcal{C}^{1,2}$, where $\mathcal{C}^{1,2}$ denotes the class of functions of two variables that are once continuously differentiable in the first argument and twice continuously differentiable in the second argument. Note that ϕ stands for both the stochastic process describing the trading activity of the large trader, and the function of time and stock. Under the assumption of a smooth strategy the effective dynamics for the underlying asset are given by

$$dS_t = b(t, S)dt + S_t v(t, S) dW_t^{(S)},$$

where

$$v(t, S) = \frac{\sigma}{1 - \rho S \frac{\partial \phi}{\partial S}},$$

$$b(t, S) = \frac{\rho S}{1 - \rho S \frac{\partial \phi}{\partial S}} \left(\frac{\partial \phi}{\partial t} + \frac{1}{2} \frac{\partial^2 \phi}{\partial S^2} S^2 v^2 \right),$$

assuming $\rho S \frac{\partial \phi}{\partial S} < 1$. The volatility has changed from the BS volatility σ to the effective volatility $\frac{\sigma}{1 - \rho S \frac{\partial \phi}{\partial S}}$, due to the feedback effects of the strategy ϕ.

As discussed in Frey [22], there are two basic types of trading strategies: On the one hand, the large investor could employ a *positive feedback* strategy, i.e. a strategy with $\frac{\partial \phi}{\partial S} > 0$. That means buying the risky asset when the price is increasing, and selling when the price is declining. This reinforces the effect of rising or falling prices, since the effective volatility increases in this case. In a standard BS model one would use such a strategy to duplicate a convex payoff like a long call. On the other hand, the large trader could follow a *contrarian feedback* strategy, i.e. $\frac{\partial \phi}{\partial S} < 0$, which means buying stocks when prices fall and vice versa, thus absorbing volatility. This would be the strategy used to duplicate a concave payoff, like a short call.

In order to generalize this framework, a deterministic liquidity function $\rho(S)$ instead of constant liquidity is introduced in Frey and Patie [23], which is referred to as the DL model.

5.2.2 Stochastic Liquidity

5.2.2.1 The Model

In our model, the underlying price process is assumed to follow the SDE

$$dS_t = \mu_t S_{t-} dt + \sigma S_{t-} dW_t^{(S)} + \rho_t S_{t-} d\phi_t^{(+)}, \tag{5.1}$$

where μ_t is a fixed continuous function of time. We additionally assume that ρ_t is a continuous stochastic process with dynamics given by

$$d\rho_t = \eta(t,\rho)dt + \nu(t,\rho)dW_t^{(\rho)}, \tag{5.2}$$

where $dW_t^{(S)} \bullet dW_t^{(\rho)} = \gamma\, dt$ for an arbitrary, but constant γ.

An intuitive specification for ρ could for instance be a mean reversion process, since we may assume a natural long-run level θ of liquidity in the market, i.e. we think of a specification for the drift given by $\eta(t,\rho) = \kappa(\theta - \rho)$, where κ denotes the speed of mean reversion. We further assume that the process stays strictly positive for $\rho_0 > 0$. This restricts additionally the choices for the volatility function $\nu(t,\rho)$. For instance, one might use functions of the type $\nu(t,\rho) = \nu\sqrt{\rho}$. To guarantee a strictly positive stock price for every point in time, we assume that ρ is bounded from above, i.e. $\rho \leq \tilde{\rho}$ such that $\tilde{\rho}\Delta\phi^{(+)} > -1$. This may be achieved either by reflecting the process ρ at a barrier $\tilde{\rho}$ or by restricting the choices of drift and diffusion coefficients to satisfy the conditions of a bounded Feller diffusion, as discussed in Breiman [10].

Note that in our setup we add a deterministic drift term μ_t to the original stock price dynamics and relax the assumption of a zero interest rate for the MMA. Thus, besides the underlying asset S, there exists the MMA which is assumed to be perfectly liquid, earning the deterministic interest rate $r_t > 0$. This does not change the derivations, but an interest rate different from zero becomes important in the second example discussed in 5.3.2.

Extending the CL model, we assume that the number of shares ϕ held by the large investor may not only depend on S and t, but also on the liquidity ρ. A plausible dependence of ϕ on ρ is depicted in the following scenario: The more illiquid the market, the fewer shares the large trader will hold due to external or internal regulations, no matter whether a positive or contrarian feedback strategy is considered. Thus, a reasonable choice would be a decreasing absolute ϕ-value with respect to ρ (for all S). The impact of S on ϕ corresponds to the positive and the contrarian feedback strategy as the two basic types of trading strategies considered: The first is an increasing function ϕ of S for all ρ, the latter a decreasing function ϕ of S for all ρ.

To expound on the differences between the CL and the SL model, consider the following scenario: The large trader is assumed to hold ϕ_0 shares of the asset. We now look at the basic strategy, that the large trader sells all his holdings of the asset when the price falls below a certain level $\bar{s}$. Selling ϕ_0 shares all at once in the BS model has no price impact at all, since the market is assumed to be perfectly liquid. Let us now consider the liquidity models. We associate the stopping time τ with the first time when a certain price bound $\bar{s}$ is undershot. The relative downward jump in the stock price process is given by $1 - \rho_\tau \phi_0$ so that we obtain a new stock price of $S_\tau = \bar{s}(1 - \rho_\tau \phi_0)$. The price discount due to illiquidity is given by $\bar{s}\rho_\tau \phi_0$ per unit, such that the total loss is $\phi_0^2 \bar{s} \rho_\tau$. The price impact in the CL model is known in advance, since $\rho_\tau \equiv$ const. such that the strategy of the large trader could be adjusted to achieve the desired amount of $\phi_0 \bar{s}$. This stands in contrast to the SL model, where liquidity changes randomly and the price impact is therefore stochastic. This example is further discussed in the second part of this chapter, providing a liquidity insurance that compensates for the loss due to illiquidity.

As mentioned before, the large trader may be forced to sell due to external or internal regulations not only when prices fall, but also when liquidity decreases. Similarly to the scenario depicted above, the large trader will incur a loss due to illiquidity in this case.

5.2.2.2 Stock Price Dynamics with Feedback Effects

We now derive and analyze the effective price process for the SL model. Rewriting the dynamics given in (5.1) and (5.2) using the Cholesky decomposition we obtain

$$dS_t = \mu_t S_{t-} dt + \sigma S_{t-} dW_t + \rho_t S_{t-} d\phi_t^{(+)}, \tag{5.3}$$

$$d\rho_t = \eta(t,\rho)dt + \nu(t,\rho)\gamma dW_t + \nu(t,\rho)\sqrt{1-\gamma^2} d\bar{W}_t, \tag{5.4}$$

with a two-dimensional standard Brownian motion $(W, \bar{W})$. We now consider a smooth trading strategy $\phi := \phi(t, S, \rho) \in \mathcal{C}^{1,2,2}$, where $\mathcal{C}^{1,2,2}$ denotes the class of functions of three variables, that are once continuously differentiable in the first argument and twice continuously differentiable in the second and in the third argument. An application of Itô's formula leads to the effective asset price dynamics stated in:

Theorem 5.1. (Effective Dynamics of the Underlying Variables) *Suppose the trading strategy of the large trader is given by $\phi(t, S, \rho) \in \mathcal{C}^{1,2,2}$. Under the assumption of $\rho S \frac{\partial \phi}{\partial S} < 1$ for all points in time, the solution to the system of stochastic differential equations (5.3) and (5.4) satisfies*

$$dS_t = b(t,S,\rho)S dt + v(t,S,\rho)S dW_t + \bar{v}(t,S,\rho)S d\bar{W}_t, \tag{5.5}$$

$$d\rho_t = \eta(t,\rho)dt + \nu(t,\rho)\gamma dW_t + \nu(t,\rho)\sqrt{1-\gamma^2} d\bar{W}_t, \tag{5.6}$$

where

$$v(t,S,\rho) = \frac{\sigma}{1-\rho S \frac{\partial \phi}{\partial S}} + \gamma \frac{\nu \rho \frac{\partial \phi}{\partial \rho}}{1-\rho S \frac{\partial \phi}{\partial S}}, \tag{5.7}$$

$$\bar{v}(t,S,\rho) = \sqrt{1-\gamma^2} \frac{\nu \rho \frac{\partial \phi}{\partial \rho}}{1-\rho S \frac{\partial \phi}{\partial S}}, \tag{5.8}$$

$$b(t,S,\rho) = \frac{\rho}{1-\rho S \frac{\partial \phi}{\partial S}} \left[\frac{\mu}{\rho} + \frac{\partial \phi}{\partial t} + \eta \frac{\partial \phi}{\partial \rho} + \frac{1}{2}\frac{\partial^2 \phi}{\partial \rho^2}\nu^2 + \frac{1}{2}\frac{\partial^2 \phi}{\partial S^2} S^2 (v^2 + \bar{v}^2) + \nu S \frac{\partial^2 \phi}{\partial S \partial \rho}\left(\gamma v + \sqrt{1-\gamma^2}\bar{v}\right) \right].$$

The total instantaneous volatility v_{tot} is equal to

$$v_{\text{tot}}(t,S,\rho) = \sqrt{v^2 + \bar{v}^2} = \frac{\sqrt{\sigma^2 + \nu^2\rho^2 \left(\frac{\partial \phi}{\partial S}\right)^2 + 2\gamma\sigma\nu\rho\frac{\partial \phi}{\partial \rho}}}{1-\rho S \frac{\partial \phi}{\partial S}}, \tag{5.9}$$

whereas the instantaneous correlation $\mathrm{corr}_{S,\rho}$ *between the processes* S *and* ρ *is given by*

$$\mathrm{corr}_{S,\rho}(t,S,\rho) = \frac{\gamma v + \sqrt{1-\gamma^2}\bar{v}}{\sqrt{v^2+\bar{v}^2}} = \frac{\nu\rho\frac{\partial\phi}{\partial\rho} + \gamma\sigma}{\sqrt{\sigma^2 + \nu^2\rho^2\left(\frac{\partial\phi}{\partial\rho}\right)^2 + 2\gamma\sigma\nu\rho\frac{\partial\phi}{\partial\rho}}}\,. \tag{5.10}$$

The effective dynamics of the stochastic liquidity model belong to the class of diffusion models. The proof of this result is straightforward.

Proof: First, we have to note that S has continuous paths since $\phi(t,S,\rho) \in \mathcal{C}^{1,2,2}$. In this case it holds that $S_{t-} = S_t$ and $\phi^{(+)} = \phi$. Thus, we can omit the subscript $t-$ and the superscript $(+)$. To simplify the notation, we furthermore omit the arguments of the functions under consideration in the following calculations. Applying Itô to $\phi(t,S,\rho)$ yields

$$\begin{aligned} d\phi_t = {} & \frac{\partial\phi}{\partial t}dt + \frac{\partial\phi}{\partial S}dS_t + \frac{\partial\phi}{\partial\rho}d\rho_t + \frac{\partial^2\phi}{\partial S\partial\rho}(dS_t \bullet d\rho_t) \\ & + \frac{1}{2}\left(\frac{\partial^2\phi}{\partial S^2}(dS_t \bullet dS_t) + \frac{\partial^2\phi}{\partial\rho^2}\nu^2 dt\right). \end{aligned}$$

Plugging this into (5.3), we get

$$\begin{aligned} dS_t = {} & \left(\sigma + \rho\frac{\partial\phi}{\partial\rho}\nu\gamma\right)SdW_t + \rho\frac{\partial\phi}{\partial\rho}\nu\sqrt{1-\gamma^2}Sd\bar{W}_t + \left(\mu S + \rho S\frac{\partial\phi}{\partial t}\right)dt \\ & + \rho S\left(\frac{\partial\phi}{\partial S}dS_t + \frac{\partial\phi}{\partial\rho}\eta dt + \frac{\partial^2\phi}{\partial S\partial\rho}(dS_t \bullet d\rho_t)\right. \\ & \left. + \frac{1}{2}\left(\frac{\partial^2\phi}{\partial S^2}(dS_t \bullet dS_t) + \frac{\partial^2\phi}{\partial\rho^2}\nu^2 dt\right)\right). \end{aligned}$$

This is equivalent to

$$\begin{aligned} dS_t = {} & \frac{1}{1-\rho S\frac{\partial\phi}{\partial S}}\left(\left(\sigma + \rho\frac{\partial\phi}{\partial\rho}\nu\gamma\right)SdW_t + \rho\frac{\partial\phi}{\partial\rho}\nu\sqrt{1-\gamma^2}Sd\bar{W}_t\right. \\ & + \rho S\left[\left(\frac{\mu}{\rho} + \frac{\partial\phi}{\partial t} + \eta\frac{\partial\phi}{\partial\rho} + \frac{1}{2}\frac{\partial^2\phi}{\partial\rho^2}\nu^2\right)dt + \frac{1}{2}\frac{\partial^2\phi}{\partial S^2}(dS_t \bullet dS_t)\right. \\ & \left.\left. + \frac{\partial^2\phi}{\partial S\partial\rho}(dS_t \bullet d\rho_t)\right]\right) \end{aligned}$$

assuming $\rho S\frac{\partial\phi}{\partial S} < 1$. Using the trial solution

$$dS_t = b(t,S,\rho)Sdt + v(t,S,\rho)SdW_t + \bar{v}(t,S,\rho)Sd\bar{W}_t$$

and comparing coefficients leads to

$$
\begin{aligned}
v(t,S,\rho) &= \frac{\sigma + \rho\frac{\partial\phi}{\partial\rho}\nu\gamma}{1-\rho S\frac{\partial\phi}{\partial S}}, \\
\bar{v}(t,S,\rho) &= \frac{\nu\rho\frac{\partial\phi}{\partial\rho}\sqrt{1-\gamma^2}}{1-\rho S\frac{\partial\phi}{\partial S}}, \\
b(t,S,\rho)dt &= \frac{\rho}{1-\rho S\frac{\partial\phi}{\partial S}}\Bigg[\left(\frac{\mu}{\rho}+\frac{\partial\phi}{\partial t}+\eta\frac{\partial\phi}{\partial\rho}+\frac{1}{2}\frac{\partial^2\phi}{\partial\rho^2}\nu^2\right)dt \\
&\qquad + \frac{1}{2}\frac{\partial^2\phi}{\partial S^2}(dS_t\bullet dS_t)+\frac{\partial^2\phi}{\partial S\partial\rho}(dS_t\bullet d\rho_t)\Bigg] \\
&= \frac{\rho}{1-\rho S\frac{\partial\phi}{\partial S}}\Bigg[\left(\frac{\mu}{\rho}+\frac{\partial\phi}{\partial t}+\eta\frac{\partial\phi}{\partial\rho}+\frac{1}{2}\frac{\partial^2\phi}{\partial\rho^2}\nu^2\right)+\frac{1}{2}\frac{\partial^2\phi}{\partial S^2}S^2\times \\
&\qquad \frac{\sigma^2+\rho^2\nu^2\left(\frac{\partial\phi}{\partial\rho}\right)^2+2\gamma\rho\sigma\nu\frac{\partial\phi}{\partial\rho}}{(1-\rho S\frac{\partial\phi}{\partial S})^2}+\nu\frac{\partial^2\phi}{\partial S\partial\rho}S\frac{\nu\rho\frac{\partial\phi}{\partial\rho}+\gamma\sigma}{1-\rho S\frac{\partial\phi}{\partial\rho}}\Bigg]dt,
\end{aligned}
$$

since

$$
\begin{aligned}
dS_t\bullet dS_t &= S^2(v^2+\bar{v}^2)dt \\
&\stackrel{!}{=} S^2\frac{\sigma^2+\rho^2\nu^2\left(\frac{\partial\phi}{\partial\rho}\right)^2+2\gamma\rho\sigma\nu\frac{\partial\phi}{\partial\rho}}{(1-\rho S\frac{\partial\phi}{\partial S})^2}dt \\
dS_t\bullet d\rho_t &= S\nu(\gamma v+\sqrt{1-\gamma^2}\bar{v})dt \\
&\stackrel{!}{=} S\nu\frac{\nu\rho\frac{\partial\phi}{\partial\rho}+\gamma\sigma}{1-\rho S\frac{\partial\phi}{\partial S}}dt.
\end{aligned}
$$

Heuristically, the correlation between S and ρ is computed by $\frac{dS_t\bullet d\rho_t}{\nu S v_{\text{tot}}dt}$.

Since the volatility coefficients are stochastic, our model setup does not only fit into the class of diffusion models, but it belongs in particular to the SV class of models.

For $\rho\equiv 0$ or $\phi\equiv$ constant we are in the classical BS scenario with drift μ_t. If $\rho\neq 0$ and ϕ not being a function of time only, the trading strategy of the large trader has an effect on the instantaneous volatilities v and $\bar{v}$, as well as on the total volatility and the correlation between the processes S and ρ.

Consider first the special case that liquidity has no impact on the trading strategy of the large investor, i.e. $\frac{\partial\phi}{\partial\rho}\equiv 0$. Then, $\bar{v}$ and the second summand of v will vanish in (5.7). This means that the effective dynamics of the underlying asset are close to the CL model, where feedback effects are only incorporated in the term $\rho S\frac{\partial\phi}{\partial S}$. In this case, the correlation of the processes ρ and S is equal to γ, which is exactly the correlation between the two increments of the Brownian motion. This is in general not true as we will see below.

The results are much more complex for $\frac{\partial \phi}{\partial \rho} \neq 0$. Assuming $\rho > 0$, we now analyze the properties of the derivative $\frac{\partial \phi}{\partial \rho}$, in order to discuss how a change in liquidity influences the trading strategy of the large investor. The more illiquid the market becomes, the more eager the large trader is to close his position, which means to sell everything he is long or to buy back everything he is short. In the first case ϕ is monotonically decreasing in ρ, starting with a positive ϕ. In the latter case ϕ is monotonically increasing in ρ starting with a negative ϕ. Thus, ϕ is approaching zero in absolute value as ρ tends to infinity. In the following, we assume a positive ϕ so that $\frac{\partial \phi}{\partial \rho}$ is negative, no matter if a positive feedback or a contrarian trading strategy is considered. This especially yields $\bar{v} < 0$.

In order to compare the CL and DL setup with the SL model, we analyze the volatility and correlation structure for different specifications of the respective liquidity-related parameters. An overview is given in Table 5.1. For convenience, the partial derivatives are denoted by subscripts.

Model	ρ	γ	v	$\bar{v}$	v_{tot}	$\text{corr}_{S,\rho}$
BS	0	0	σ	0	v	0
CL	const.	0	$\frac{\sigma}{1-\rho S \phi_S}$	0	v	0
DL	$\rho(S)$	1	$\frac{\sigma}{1-\rho S \phi_S}$	0	v	1
SL	stoch.	0	$\frac{\sigma}{1-\rho S \phi_S}$	$\frac{\nu \rho \phi_\rho}{1-\rho S \phi_S}$	$\sqrt{v^2 + \bar{v}^2} > v$	$\frac{\bar{v}}{\sqrt{v^2+\bar{v}^2}} < 0$
SL	stoch.	$\neq 0$	$\frac{\sigma+\gamma \nu \rho \phi_\rho}{1-\rho S \phi_S}$	$\frac{\sqrt{1-\gamma^2}\nu \rho \phi_\rho}{1-\rho S \phi_S}$	$\sqrt{v^2 + \bar{v}^2}$	$\frac{\sqrt{1-\gamma^2}v+\gamma \bar{v}}{\sqrt{v^2+\bar{v}^2}}$

Table 5.1. Liquidity-Related Parameters in the Different Models

We start with the CL model as the first extension of BS with constant ρ (i.e. $d\rho \equiv 0$). This yields $\nu \equiv 0, \bar{v} \equiv 0$, $\text{corr}_{S,\rho} \equiv 0$, and

$$v \equiv v_{\mathrm{tot}} = \frac{\sigma}{1 - \rho S \frac{\partial \phi}{\partial S}}.$$

Especially all terms containing $\frac{\partial \phi}{\partial \rho}$ vanish so that the dependence of the strategy on ρ is of no interest for the effective stock price dynamics. A deterministic liquidity function $\rho(S)$ instead of constant liquidity is considered in the DL model. Since ρ is a deterministic function of S in this case, the dynamics of ρ are only driven by the first component W of the two-dimensional Brownian motion, implying $\bar{v} \equiv 0$ and $\gamma = 1$. Since ϕ only depends on S in the DL setup, the volatility is the same as in the CL model. The facts that $\bar{v} \equiv 0$ and $\gamma \equiv 1$ yield $\mathrm{corr}_{S,\rho} \equiv 1$ in this case, which shows the perfect correlation of the increments of the stock price process and the liquidity process. This is inherited from the perfect correlation of the components of the Brownian motion. Thus, the DL model is a special case of our setup,[1] which is obtained for $\gamma = 1$.

We now consider the SL model: Let us first look at the correlation structure. Surprisingly, $\gamma \equiv 0$ in (5.10) does not imply that the increments of the effective stock price process and the liquidity process are uncorrelated. Indeed, in this setup $\mathrm{corr}_{S,\rho} = 0$ cannot be obtained for a deterministic choice of γ. In general, $\mathrm{corr}_{S,\rho}$ changes randomly over time in our setting due to the stochastic ρ and $\frac{\partial \phi}{\partial \rho}$. For $\gamma = 0$ we still obtain a negative correlation between S and ρ, induced by $\phi_\rho < 0$:

$$\mathrm{corr}_{S,\rho} = \frac{\nu \rho \frac{\partial \phi}{\partial \rho}}{\sqrt{\sigma^2 + \nu^2 \rho^2 \left(\frac{\partial \phi}{\partial \rho}\right)^2}} < 0.$$

This term ultimately models liquidity feedback effects, i.e. the impact of a trading strategy depending explicitly on liquidity. It vanishes for a trading strategy independent of ρ, i.e. for $\frac{\partial \phi}{\partial \rho} \equiv 0$. The negative value of $\mathrm{corr}_{S,\rho}$ due to liquidity effects can be heuristically interpreted in the following way: If liquidity is low, the trader will be forced to close his position, which will cause the stock price to drop.

For $\gamma \neq 0$ the numerator of $\mathrm{corr}_{S,\rho}$ in (5.10) carries an additional summand $\gamma\sigma$. This causes $\mathrm{corr}_{S,\rho}$ to be an increasing function of γ. For $\frac{\partial \phi}{\partial \rho} \equiv 0$ this

[1] To be precise, for non-monotonic functions $\rho(S)$ the specification of the dynamics of ρ must be slightly more general such that the drift and diffusion coefficients are allowed to depend additionally on S.

correlation is equal to the instantaneous correlation γ between the increments of the components of the Brownian motion. Thus, the difference between γ and $\mathrm{corr}_{S,\rho}$ is a result of the liquidity feedback effect, exclusively.

Now we discuss the impact of SL on volatility: First, consider the case $\gamma = 0$. Then, v is the same as in the CL model, but in our framework we have the additional volatility parameter

$$\bar{v} = \frac{\nu\rho\frac{\partial\phi}{\partial\rho}}{1 - \rho S\frac{\partial\phi}{\partial S}} < 0,$$

corresponding to the second source of risk $\bar{W}$ that contributes to total volatility. The variable $\bar{v}$ also incorporates the liquidity feedback effect. It contributes to the total volatility, if and only if the trading strategy actually varies with changing liquidity, i.e. for $\frac{\partial\phi}{\partial\rho} \neq 0$. Then the total instantaneous volatility

$$v_{\mathrm{tot}} = \frac{\sqrt{\sigma^2 + \nu^2\rho^2\left(\frac{\partial\phi}{\partial\rho}\right)^2}}{1 - \rho S\frac{\partial\phi}{\partial S}}$$

increases, compared with the adjusted volatility in the CL setup. Here we see a key result of our approach: For $\gamma = 0$ the total volatility of the SL model is greater than the volatility in the CL framework, no matter if a positive or a contrarian strategy is used.

For $\gamma \neq 0$ it is more difficult to analyze the characteristics of the model. The direction of the changes in total volatility from (5.9) depends on the sign of γ. The parameter v carries an additional summand describing the liquidity feedback effect, given by $\frac{\partial\phi}{\partial\rho}$, and $\bar{v}$ is reduced in absolute value due to $\gamma \neq 0$, as we can see from (5.7) and (5.8). The total volatility in (5.9) is a decreasing function of γ, since $\frac{\partial\phi}{\partial\rho} < 0$. Thus, one implication of our analysis is that for all $\gamma \leq 0$ the total volatility is greater than the volatility in the CL setup for any smooth strategy with $\frac{\partial\phi}{\partial\rho} \neq 0$. This follows from the fact that the trading strategy also reacts to changes in liquidity, which has an additional impact on the asset price dynamics. Only for large positive values of γ can the total volatility be lower than the volatility in the CL model, and this is the case if and only if

$$1 \geq \gamma > \frac{1}{2\sigma}\nu\rho\left|\frac{\partial\phi}{\partial\rho}\right|. \tag{5.11}$$

For a trading strategy ϕ independent of ρ, the volatility in the SL model equals the volatility in the CL model, again reflecting the lack of the liquidity feedback effect in this case.

Finally, we look at the two basic types of trading strategies discussed in Frey [22] with respect to S. A positive feedback strategy, i.e. $\frac{\partial\phi}{\partial S} > 0$, leads to

$$v_{\text{tot}} > \sqrt{\sigma^2 + \rho^2\nu^2\left(\frac{\partial\phi}{\partial\rho}\right)^2 + 2\gamma\sigma\nu\rho\frac{\partial\phi}{\partial\rho}}.$$

The expression on the right-hand side is greater than or equal to σ, if and only if

$$\gamma \leq \frac{1}{2\sigma}\nu\rho\left|\frac{\partial\phi}{\partial\rho}\right|. \tag{5.12}$$

This especially implies the following result: The instantaneous volatility v_{tot} in the SL model is greater than the BS volatility σ for $\gamma \leq 0$, considering a positive feedback strategy. This is similar to the result derived in the CL model.

A contrarian feedback strategy (i.e. $\frac{\partial\phi}{\partial S} < 0$) implies that the total volatility satisfies

$$v_{\text{tot}} < \sqrt{\sigma^2 + \rho^2\nu^2\left(\frac{\partial\phi}{\partial\rho}\right)^2 + 2\gamma\sigma\nu\rho\frac{\partial\phi}{\partial\rho}}.$$

The expression on the right-hand side is less than σ, if and only if (5.11) holds. Thus, for non-positive values of γ the instantaneous volatility in the SL model is not lower than the BS volatility, considering a contrarian feedback strategy. This stands in contrast to the result derived in the CL model.

Now we look at further implications of our model. The results of Chap. 3 can be applied here, since the extended liquidity model has turned out to fit into a general diffusion framework. We have not analyzed the drift under the physical measure, as it is not relevant for no-arbitrage pricing. We next consider the drift restriction, the change of measure, including a vector of market prices of risk, and the stock price process under a risk-neutral measure in this model setup.

5.2.2.3 Risk-Neutral Dynamics

Based on the results presented in 3.4.1.1, the drift restriction in the SL model is given by

$$b(t, S, \rho) - r_t = \lambda_t v_t + \bar{\lambda}_t \bar{v}_t. \tag{5.13}$$

Since the market is incomplete, there is one degree of freedom, and the market price of risk vector $\Lambda = (\lambda, \bar{\lambda})$ for the sources of risk $(W, \bar{W})$ is not uniquely determined. This situation is similar to the SV example discussed in Sect. 3.5. For $\bar{v} = 0$, i.e. if $\frac{\partial \phi}{\partial \rho} \equiv 0$ the first component of the market price of risk is given by

$$\lambda_t = \frac{b(t, S, \rho) - r_t}{v_t}, \tag{5.14}$$

and $\bar{\lambda}_t$ is an arbitrary predictable process with respect to $\mathcal{F}_t$. For constant liquidity the market is no longer incomplete, and the market price of risk is uniquely determined by (5.14). For stochastic liquidity the change of measure to a risk-neutral measure $\hat{P}$ is given by Girsanov's theorem, as discussed in Proposition 3.1, using a market price of risk vector Λ given above:

$$\begin{aligned} d\hat{W}_t &= \lambda_t dt + dW_t \\ d\hat{\bar{W}}_t &= \bar{\lambda}_t dt + d\bar{W}_t. \end{aligned}$$

Since the market is incomplete, $\hat{P}$ is not unique. The resulting dynamics under $\hat{P}$ are

$$dS_t = r_t S_t dt + v(t, S, \rho) S_t d\hat{W}_t + \bar{v}(t, S, \rho) S_t d\hat{\bar{W}}_t \tag{5.15}$$

$$d\rho_t = \eta^*(t, \rho) dt + \nu(t, \rho) \gamma d\hat{W}_t + \nu(t, \rho) \sqrt{1 - \gamma^2} d\hat{\bar{W}}_t, \tag{5.16}$$

using the drift restriction in (5.13) for the underlying asset S. The drift coefficient $\eta^*(t, \rho)$ under $\hat{P}$ for the liquidity process is given by

$$\eta^*(t, \rho) = \eta(t, \rho) - \gamma \nu(t, \rho) \lambda_t - \sqrt{1 - \gamma^2} \nu(t, \rho) \bar{\lambda}_t. \tag{5.17}$$

5.3 Examples

In this section we present two illustrations of our model. They highlight some features which are exclusive to the SL setup. First, we will look at two types of trading strategies, independent of time, and their impact on the stock price processes to illustrate the theoretical findings. Second, we will develop an insurance contract, offering protection against the illiquidity discount when large positions are liquidated at once.

5.3.1 Numerical Analysis of the Effective Stock Price Dynamics for Two Trading Strategies

In the first example we present some simulation-based results for the stock price processes under the physical measure in the two models: We consider the SL model, while the BS model (i.e. $\rho \equiv 0$ or $\phi \equiv 0$) with zero drift (i.e. $\mu \equiv 0$) serves as the benchmark. The stock price dynamics in an SL setting for both positive and contrarian feedback strategies are compared with the BS dynamics.

5.3.1.1 Typical Feedback Strategies

We propose an ad-hoc trading strategy with an economically intuitive functional form for the dependence of the strategy ϕ on the stock price S and the liquidity parameter ρ. As before, we restrict ourselves to positive holdings of the asset by the large trader, i.e. $\phi > 0$, at any point in time. The initial holdings ϕ_0 of the asset are assumed to be optimal with respect to ρ and S, in the sense that the large trader has no intention to buy or sell additional units of the asset at time $t = 0$, i.e. $\phi_0 = \phi(0, S, \rho)$.

Figure 5.1 shows the dependence of ϕ on ρ and S for a positive (left graph) and a contrarian (right graph) feedback strategy.

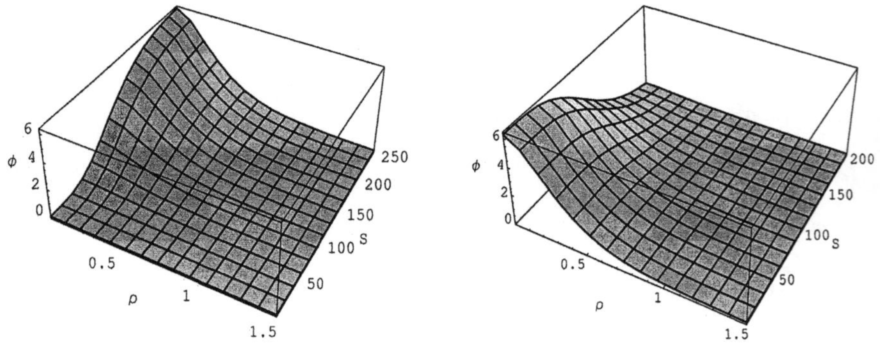

Fig. 5.1. Positive and Contrarian Feedback Strategy

The graphs shown in the two figures illustrate the scenario which has already been discussed in the introduction. If liquidity drops, i.e. ρ increases, the large trader is forced to sell some shares due to external regulations.

Consider for example a mutual fund that is allowed to invest only in stocks that are part of a highly liquid index. When the stock becomes illiquid and therefore no longer belongs to the index, the fund has to close its position in this asset. Therefore, ϕ is assumed to be monotonically decreasing in ρ for all S and for any feedback strategy, i.e. the derivative of ϕ with respect to ρ is negative. It seems reasonable to assume that for small values of ρ the value of $\frac{\partial \phi}{\partial \rho}$ is also small in absolute value, since changes in small values of ρ have only small effects on the market depth of the asset. If ρ increases, i.e. liquidity declines, the trader wants to get rid of his assets, thus $\frac{\partial \phi}{\partial \rho}$ will increase in absolute values. However, for large values of ρ the absolute value of $\frac{\partial \phi}{\partial \rho}$ will again decrease and approach zero, since the large trader has already sold almost all of his holdings in the stock.

In order to characterize the relationship between ϕ and S, we have to distinguish between the positive feedback and the contrarian feedback strategy.

In the first (second) case the large trader buys (sells) assets as the stock price increases and sells (buys) when the stock price declines. Thus, ϕ is monotonically increasing (decreasing) in S for all ρ in the case of the positive (contrarian) feedback strategy. For very small and very large values of S the changes in the stock holdings with varying stock prices become negligible when the asset prices vary (similar to the relationship between ρ and ϕ). However, for intermediate asset prices the absolute value of ϕ_S increases, when S increases and a positive (contrarian) feedback strategy is considered.

There is a variety of functional forms for $\phi = \phi(S, \rho)$ that are able to reproduce the features described above. In order to incorporate the behavior of ϕ as a function of both S and ρ, we choose a product form separating the strategy with respect to S and ρ, i.e. $\phi(S, \rho) = a_0 \psi(S) \chi(\rho)$. A function that is able to model the depicted monotonicity behavior – as a function of the second argument for suitably chosen first argument – is the incomplete gamma function, defined by

$$\Gamma(x, z) = \int_z^\infty e^{-t} t^{x-1} dt.$$

We now propose the following representations for appropriate constants d_1, d_2, e_1, e_2:

$$\chi(\rho) = \Gamma(d_1, e_1\rho),$$

$$\psi(S) = \begin{cases} \Gamma(d_2, e_2 S) & \text{contrarian strategy,} \\ \\ \Gamma(d_2) - \Gamma(d_2, e_2 S) & \text{positive strategy,} \end{cases}$$

where $\Gamma(x) = \Gamma(x, 0)$ denotes the standard gamma function. The contrarian feedback strategy is given by

$$\phi^{(c)}(S, \rho) = a_0 \Gamma(d_1, e_1\rho)\, \Gamma(d_2, e_2 S), \tag{5.18}$$

whereas the positive feedback strategy is given by

$$\phi^{(p)}(S, \rho) = a_0 \Gamma(d_1, e_1\rho)\, [\Gamma(d_2) - \Gamma(d_2, e_2 S)]\,. \tag{5.19}$$

For the plots shown above and the simulations below, we have used the following parameters:

$$\begin{array}{lll} d_1 = 2.1 & d_2 = 6.00 & \\ e_1 = 5.5 & e_2 = 0.05 & a_0 = 0.05. \end{array}$$

Moreover, we have taken the initial values $\rho_0 = 0.05$ and $S_0 = 80$ in the simulation. The slope of both $\phi^{(c)}$ and $\phi^{(p)}$ with respect to ρ is steepest for ρ around 0.1 and steepest with respect to S for S around 120.

Note that in order to ensure that the SDE (5.5) has a solution, we have to guarantee that $\rho S \frac{\partial \phi}{\partial S} < 1$. For our choice of parameters and assuming $\rho, S \in \mathbb{R}^+$, the expression in (5.18) has a global maximum of 0.979 at $S = 120$ and $\rho = 0.306$. Since in (5.19) the partial derivative $\frac{\partial \phi^{(p)}}{\partial S}$ is negative, the condition $\rho S \frac{\partial \phi}{\partial S} < 1$ is obviously satisfied.

5.3.1.2 Parameter Specifications for the Sample Paths

To create paths of the underlying asset and liquidity a Monte Carlo simulation is used. The stochastic processes are discretized with an Euler scheme with 4000 time steps and $\Delta t = 1/360$. In order to ensure non-negativity and stationary behavior of the liquidity process we specify the dynamics for ρ in (5.2) as a square-root process with a mean reverting drift component:

$$\eta(t, \rho) = \kappa(\theta - \rho), \quad \nu(t, \rho) = \nu\sqrt{\rho}$$

(see Cox, Ingersoll, and Ross [12]). For a fixed realization of $(W_t, \bar{W}_t)$ we have plotted the dynamics of (5.3) and (5.4) for (I) BS with $\rho \equiv 0$, i.e. $\eta \equiv \nu \equiv 0$ and (II) SL with the parameter values

$$S_0 = 80.0 \qquad \kappa = 0.35 \qquad \theta = \rho_0 = 0.05$$
$$\sigma = 0.1 \qquad \mu = 0.00 \qquad \gamma = 0.00$$
$$\nu = 0.20$$

for (i) the contrarian feedback strategy, i.e. $\phi = \phi^{(c)}$ in (5.18) and for (ii) the positive feedback strategy, i.e. $\phi = \phi^{(p)}$ in (5.19). For both strategies the drift under the physical measure is random, but close to zero using these parameter values.

5.3.1.3 Positive Feedback Strategy

In Fig. 5.2 we compare the BS path with the stock price path in the SL model using a positive feedback strategy. The paths shown are typical scenarios with respect to the physical measure occurring for a long time horizon.

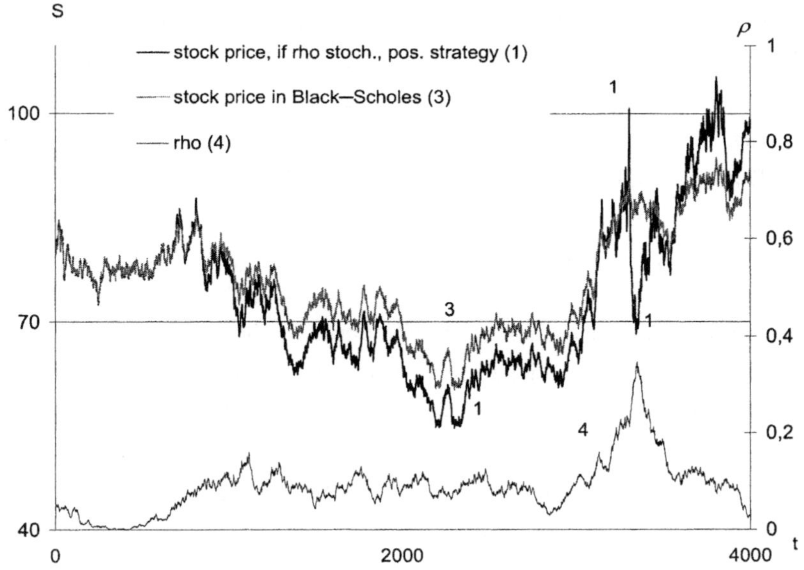

Fig. 5.2. Sample Paths for the Stock Price in the BS Model and the SL Setup if the Large Trader Follows a Positive Feedback Strategy

The left axis shows the values for the stock price dynamics. Path 3 corresponds to the BS dynamics and path 1 to the stock price dynamics with stochastic liquidity using a positive feedback strategy. The right axis corresponds to the values of the liquidity path 4. As shown theoretically (see (5.12)), it is illustrated that for our parameter choice the volatility of the stock price in the stochastic liquidity model is increased compared with BS. The SL path 1 exceeds the BS path 3 when the BS price is increasing over a longer period. Rising stock prices motivate the large investor to buy additional stocks which will cause the stock price to grow even further. The opposite effect can be noticed for a decreasing S, since in this case the large investor wants to get rid of the holdings which will accelerate the decline in the stock price. If ρ is close to zero, the SL path 1 runs close to the BS path 3, as we can observe around time step 500. These features observed for the paths in the SL model are similar when simulating analogous sample paths in the CL setup.

However, the role of the liquidity parameter ρ is more subtle. In fact it can have two different effects: First, all else equal, the trader has to sell stocks if they become more and more illiquid. Second, if ρ is very high, the stock becomes more volatile so that a large trader who follows a positive feedback strategy can cause the stock price to rise to tremendously high values in bullish markets. However, when illiquidity exceeds a certain threshold, the large trader is forced to close the position due to illiquidity and the market collapses. These features that distinguish the stochastic liquidity model from the BS and the CL model can be seen in Fig. 5.2 around time step 3,500.

5.3.1.4 Contrarian Feedback Strategy

In Fig. 5.3 we contrast stock prices simulated in the BS model to stock prices generated in the SL model using a contrarian feedback strategy. They are again typical paths under the physical measure occurring for a long time horizon.

As before, the left axis shows the values for the stock price dynamics. Path 3 corresponds to the BS dynamics and path 2 to the stock price dynamics with stochastic liquidity using a contrarian feedback strategy. Again, the right axis corresponds to the values of the liquidity path 4. The SL path 2 stays above the BS path 3 when the BS price is decreasing over a longer period, since falling stock prices motivate the large investor to buy additional stocks. This causes the stock price to drop less than in the BS model. The opposite effect can be noticed for an increasing S, since in this case the large investor wants

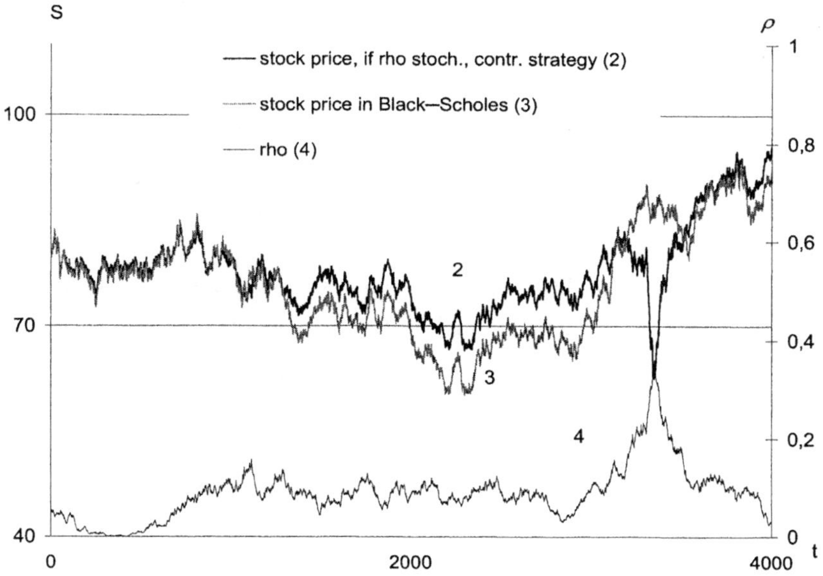

Fig. 5.3. Sample Paths for the Stock Price in the BS Model and the SL Setup if the Large Trader Follows a Contrarian Feedback Strategy

to get rid of the holdings, which reduces the increase in the stock price. The analogous sample paths in the CL setup show a behavior similar to the SL paths.

However, as shown theoretically (see (5.11)), it is – in contrast to the CL setup – no longer true in the SL model that volatility is reduced compared with BS. This feature becomes obvious around time step 3,500. If the asset becomes very illiquid the trading activities of the large trader can have a destabilizing effect on the stock price dynamics. Again, this is a unique feature of the SL setup that cannot be modeled in the CL framework.

5.3.2 Liquidity Insurance

This subsection proposes a liquidity derivative. For instance, we might think of a portfolio manager who has to limit the downside risk of a fund. In order to do so, we assume that he follows a simple stop loss strategy. When the price of the asset of interest falls below a certain level for the first time, the entire position in this security is immediately liquidated. If the market is perfectly liquid and the asset price process is continuous, the fund will always receive the stop loss limit for the asset. However, if the market becomes illiquid, the investor will receive less than the stop loss limit. As discussed before, if the degree of illiquidity does not change, the illiquidity discount is known in advance. This means that the investor can adjust the stop loss limit, so that he will always receive the desired amount for the assets. However, in the SL model the large investor faces liquidity risk.

The proposed liquidity derivative matures exactly at the point in time when the price of the underlying asset undershoots a critical level for the first time. The derivative compensates for the price difference between the asset price immediately before and immediately after the execution of the stop loss order. The general setup for the insurance contract is described, and a valuation equation for such a liquidity derivative is developed.

5.3.2.1 Specification and Pricing of the Contract

We assume in the following that ϕ represents a stop loss strategy. We denote by τ the stopping time when the underlying asset falls below a certain level $\bar{s}$ for the first time, i.e. $\tau := \inf\{t \mid S_t < \bar{s}\}$ and $S_{\tau-} = \bar{s}$. For the initial price of the underlying asset we assume $S_0 > \bar{s}$. Furthermore, we assume a constant, positive initial position $\phi_0 > 0$. Up to the hitting time τ the large trader does not trade, and at τ he sells all his assets. This implies

$$\Delta\phi_t^{(+)} = \begin{cases} 0 & \text{for } t < \tau \\ \phi_t^+ - \phi_t = -\phi_0 & \text{for } t = \tau \end{cases}$$

and

$$\begin{aligned} dS_t &= \mu_t S_t dt + \sigma S_t dW_t \\ d\rho_t &= \eta(t,\rho)dt + \nu(t,\rho)\gamma dW_t + \nu(t,\rho)\sqrt{1-\gamma^2}d\bar{W}_t \end{aligned}$$

for $t < \tau$, where $(W, \bar{W})$ is a two-dimensional Brownian motion with uncorrelated increments.

Under a risk-neutral measure $\hat{P}$ the dynamics are given by

$$dS_t = r_t S_t dt + \sigma S_t d\hat{W}_t \tag{5.20}$$

$$d\rho_t = \eta^*(t,\rho)dt + \nu(t,\rho)\gamma d\hat{W}_t + \nu(t,\rho)\sqrt{1-\gamma^2}d\hat{\bar{W}}_t \tag{5.21}$$

for $t < \tau$, where η^* denotes the risk-adjusted drift given in (5.17). This equation is valid since $\phi_t^{(+)} \equiv 0$ for all $t < \tau$ satisfies the condition in Theorem 5.1. At $\tau-$ the threshold is hit, implying a jump in the asset price. We end up with the reduced price at τ given by

$$S_\tau = \bar{s}(1 - \rho_{\tau-}\phi_0) = \bar{s}(1 - \rho_\tau \phi_0),$$

assuming a continuous process for ρ. We propose an insurance contract that covers the price difference between the reduced price at τ and the threshold $\bar{s}$ by paying

$$\zeta_\tau = \bar{s}\rho_\tau \phi_0 I(\tau \leq T)$$

at τ per unit of the underlying asset. The price ζ_0 of this insurance is given in:

Proposition 5.2. *The price at* $t = 0$ *of the contract paying*

$$\zeta_\tau = \bar{s}\rho_\tau \phi_0 I(\tau \leq T)$$

at the stopping time τ *is given by*

$$\begin{aligned} \zeta_0 &= \hat{\mathbf{E}}_0 \left[e^{-r\tau} \zeta_\tau \right] \\ &= \bar{s}\phi_0 \hat{\mathbf{E}}_0 \left[e^{-r\tau} \rho_\tau I(\tau \leq T) \right] \\ &= \bar{s}\phi_0 \int_0^T \int_0^{\tilde{\rho}} e^{-rt} \rho H(t, \rho) d\rho dt, \end{aligned} \tag{5.22}$$

where H *represents the joint density of* τ *and* ρ_τ *under* $\hat{P}$, *and* $\tilde{\rho}$ *denotes the global bound for* ρ.

This result can be obtained by risk-neutral valuation because the concept of martingale pricing does not only hold for fixed points in time, but also for stopping times (cf., inter alia, Musiela and Rutkowski [33]). For convenience, we have assumed a constant interest rate $r > 0$ in this scenario.

In general, the price of the insurance cannot be calculated explicitly. If the joint density is known, we propose numerical integration, otherwise we have to use simulation techniques. For some special cases we are able to derive explicit solutions of formula (5.22). These are presented in the following.

First of all, we need the distribution of the hitting time. The process for the underlying asset S under $\hat{P}$ up to the first hitting time τ is a geometric Brownian motion, as can be seen from (5.20). Thus,

$$\ln S_t = \log S_0 + (r - \sigma^2/2)t + \sigma \hat{W}_t^{(S)}$$

is an arithmetic Brownian motion. The distribution of the first hitting time of a Brownian motion with drift is well-known (see, e.g. Borodin and Salminen [8]). In this scenario we ask for the distribution of the first hitting time of $\ln \bar{s}$, starting at $\ln S_0 > \ln \bar{s}$. This is given by the density

$$h_{\ln S_0, \ln \bar{s}}(t) = \frac{|\ln(\bar{s}/S_0)|}{\sqrt{2\pi t^3}} \exp\left(\frac{-(\ln(\bar{s}/S_0) - (r - \sigma^2/2)t)^2}{2t} \right).$$

The specification of the diffusion process for liquidity is arbitrary. Now, we look at some special scenarios in which the price can be calculated explicitly.

In the special case of a constant or time-independent random variable for liquidity, uncorrelated with the underlying asset, (5.22) simplifies to

$$\begin{aligned} \zeta_0 &= \bar{s}\phi_0 \hat{\mathbf{E}}_0[\rho]\, \hat{\mathbf{E}}_0\left[e^{-r\tau} I(\tau \leq T)\right] \\ &= \bar{s}\phi_0 \hat{\mathbf{E}}_0[\rho] \int_0^T e^{-rt} h_{\ln S_0, \ln \bar{s}}(t)dt. \end{aligned}$$

Now we consider the special case of zero correlation between the Brownian motions, i.e. $\gamma = 0$. Then, the hitting time τ and the process ρ are independent. This means that we can calculate the expectation taking the product of the corresponding densities, i.e. we can write $H(t, \rho) = h(t)\bar{h}_t(\rho)$, where $\bar{h}_t$ denotes the density under $\hat{P}$ of the process ρ at time t. From this we get

$$\begin{aligned} \zeta_0 &= \bar{s}\phi_0 \hat{\mathbf{E}}_0\left[e^{-r\tau}\rho_\tau I(\tau \leq T)\right] \\ &= \bar{s}\phi_0 \int_0^T \left(\int_0^{\bar{\rho}} \rho \bar{h}_t(\rho) d\rho \right) e^{-rt} h_{\ln S_0, \ln \bar{s}}(t)dt. \end{aligned} \tag{5.23}$$

Now, we are free to choose an appropriate process for ρ. The only thing we need for formula (5.23) is an explicit form of the density of ρ_t. To our knowledge, the question of which specification of the liquidity process is adequate is still unanswered in the literature, and serves as a topic for further research.

A numerical example for the liquidity derivative, using order book data to calibrate the liquidity process and to compute the price of the derivative in the CL and SL setup, can be found in Esser and Mönch [21].

For correlated Brownian motions we are able to get explicit solutions for simple processes such as arithmetic and geometric Brownian motions for liquidity. However, such a choice can hardly be motivated from an economic point of view. Thus, we do not go into further details.

We finally mention a similar scenario to which our framework can also be applied: When a large trader is a call short, then he might have to buy stocks to hedge the contract when the asset price hits the strike from below for the first time. Due to illiquidity he then may have to pay a price that is higher than the stop buy limit. Thus, he may ask for a contract that gives him compensation for the higher price incurred. This example can be modeled analogously.

5.3.2.2 Alternative Scenario

As already pointed out, another relevant scenario might be the following: The large investor may be forced to sell all his holdings due to a highly illiquid market, i.e. when the liquidity process ρ hits a certain bound $\bar{\rho}$ from below where $\tilde{\rho} > \bar{\rho} > \rho_0 > 0$. Recall that $\tilde{\rho}$ denotes the global bound for the SL process.

Let $\bar{\tau}$ denote the first hitting time of $\bar{\rho}$, i.e. $\bar{\tau} := \inf\{t \,|\, \rho_t > \bar{\rho}\}$ yielding $\rho_{\tau-} = \bar{\rho}$. Then, the asset price drops, i.e.

$$S_{\bar{\tau}} = S_{\bar{\tau}-}(1 - \bar{\rho}\phi_0),$$

and the large trader incurs a loss of $S_{\bar{\tau}-}\bar{\rho}\phi_0$ per unit of the underlying asset. To set up a contract that insures against this type of loss, we have to proceed analogously as before.

The stop loss order of the large trader again is given by

$$\Delta\phi^{(+)} = \begin{cases} 0 \text{ for } t < \bar{\tau}, \\ -\phi_0 \text{ for } t = \bar{\tau}. \end{cases}$$

The insurance has to pay

$$\bar{\zeta}_{\bar{\tau}} = S_{\bar{\tau}-}\bar{\rho}\phi_0 I(\bar{\tau} \leq T)$$

at $\bar{\tau}$ to cover the price difference. The price of this insurance is given in:

Proposition 5.3. *The price at $t = 0$ of the insurance paying*

$$\bar{\zeta}_{\bar{\tau}} = S_{\bar{\tau}-}\bar{\rho}\phi_0 I(\bar{\tau} \leq T)$$

at the stopping time $\bar{\tau}$ is given by

$$
\begin{aligned}
\bar{\zeta}_0 &= \hat{\mathbf{E}}_0 \left[e^{-r\bar{\tau}} \bar{\zeta}_{\bar{\tau}} \right] \\
&= \bar{\rho}\phi_0 \hat{\mathbf{E}}_0 \left[e^{-r\bar{\tau}} S_{\bar{\tau}-} I(\bar{\tau} \leq T) \right] \\
&= \bar{\rho}\phi_0 \int_0^T \int_0^\infty e^{-rt} s \bar{H}(s,t) ds\, dt,
\end{aligned}
$$

where $\bar{H}$ denotes the joint density of $S_{\bar{\tau}-}$ and $\bar{\tau}$ under $\hat{P}$.

This result is similar to the previous one with the roles of S and ρ changed. This scenario might be of interest for risk management when a trader has to hold a highly liquid portfolio and could be forced to sell due to illiquidity.

5.4 Conclusion

In this chapter we have discussed the concept of liquidity as the sensitivity of asset prices to large trading volumes. We have introduced a new model, in fact the most natural model, including a stochastic process for liquidity, thereby extending a model by Frey [22] with constant liquidity. Furthermore, we have discussed the impact of the trading strategy of a large investor on the asset prices in the SL setup. It has been shown that in the case of a smooth trading strategy the asset price process follows a diffusion process with two sources of risk, namely asset price risk and liquidity risk. The innovative features in our model include a stochastic market depth and the modeling of liquidity feedback effects. The latter refers to the impact of changing liquidity on the trading strategy and furthermore on the asset price.

We have analyzed two types of trading strategies theoretically and numerically and derived the following results: For positive feedback strategies, volatility is increased compared with BS for non-positively correlated Brownian motions, but it is no longer true in general that contrarian feedback strategies reduce volatility as in the CL model. When the asset is highly illiquid, the trading activity of a large trader can have a destabilizing effect, no matter whether a positive or a contrarian feedback strategy is used. We have discussed situations where asset price dynamics are dominated by high illiquidity, since in this case the price process is much more sensitive to trading actions of the large trader. Furthermore, we have derived the following: For every smooth trading strategy the total volatility is increased compared with the CL volatility for a sensible correlation parameter. We have presented another illustration of the SL model: an insurance contract paying the difference between a stop loss limit and the execution price. This financial innovation can, for instance, be used for risk management if a fund manager has to sell securities in falling markets or when liquidity declines.

6

Summary and Outlook

In this chapter we summarize the key results and propose several topics for further research. The first part of this thesis has considered fundamental aspects concerning martingale pricing. The main contributions of this part are the following:

- First, we have developed a convenient form of the pricing equation for a general payoff and discussed under what circumstances the changed measure corresponds to a numeraire. In a complete market every measure, equivalent to the physical measure, is an EMM with respect to an attainable numeraire. In contrast to that, in an incomplete market the equivalent measure is an EMM with respect to an attainable numeraire, if and only if the density with respect to a risk-neutral measure is attainable.
- Second, we have investigated a structural comparison between continuous and discrete models, highlighting the similarities and main differences. We have seen that the techniques used in both setups are similar. However, there are significant differences regarding the change of measure and the properties of the Brownian motion compared with the basis martingales. We conclude that this in fact emphasizes the special nature of diffusion models.

There are various directions in which the comparison could be extended. One is to include jump (diffusion) models as a hybrid type, modeling both continuous and discrete behaviour of the underlying processes. Another aspect of the comparison between discrete and continuous setups could involve an equilibrium approach. Especially for incomplete markets, this is a promising field, being one possibility to specify the non-unique EMM founded on economic grounds. In this context, embedding the SV and the SL model as two examples into an equilibrium approach is of special interest.

The second part of the thesis has discussed two applications to incomplete markets, building on the theory of the first part. There are two key results to emphasize:

- The first is the derivation of the pricing equation for power options in SV models. The approach shown here can be extended to other types of exotic options with non-piecewise-linear terminal payoffs. An open question in this context is to find appropriate hedging strategies for non-attainable claims in an SV setting. This applies in particular to the power option which is not attainable in the SV model. One result based on the findings of the paper by Branger, Esser, and Schlag [9] is that there is no superhedge for the power option with an exponent greater than one. This might be one reason why such exotic contracts are usually capped in practice. Thus, the applicability of hedging strategies in incomplete markets to power options could serve as a further research topic.
- The second innovative contribution is the modeling of an SL setup as an extension of the CL model by Frey [22], introducing a stochastic process for liquidity. We have shown several implications of this setup: Introducing a second source of risk leads to a model with stochastic volatility. Since the trading strategy additionally depends on liquidity, there is a liquidity feedback effect which has an impact on the asset price dynamics. We have further discussed a numerical example, comparing the SL model with the BS setup, distinguishing between two types of trading strategies. As another example we have developed a liquidity insurance. This contract could be further analyzed empirically using order book data, which is done in Esser and Mönch [21], as an example to illustrate the magnitude of the illiquidity discounts in the different settings.

Moreover, there are various aspects which might serve for empirical research within the playground of stochastic liquidity. There is need for empirical evidence whether a mean reversion process for liquidity, which seems intuitive from an economic point of view, is indeed an appropriate choice.

A

Power Options in Stochastic Volatility Models

A.1 Calculations of the Characteristic Functions

Using the explicit form of $X_T \equiv \ln(S_T)$ yields

$$
\begin{aligned}
\hat{\chi}_{\ln S_T,t}(k) &= \hat{\mathbf{E}}_t\Big[\exp(ikX_T)\Big]\\
&= \hat{\mathbf{E}}_t\left[\exp\left(ik\left(X_t + r(T-t) - \int_t^T \frac{\sigma_u^2}{2}\,du + \int_t^T \sigma_u d\hat{W}_u^S\right)\right)\right]\\
&= e^{ik(r(T-t)+X_t)}\hat{\mathbf{E}}_t\left[\exp\left(ik\left(-\int_t^T \frac{\sigma_u^2}{2}du + \int_t^T \sigma_u d\hat{W}_u^S\right)\right)\right],\\
\chi^{(a)}_{\ln S_T,t}(k) &= \frac{1}{\hat{\mathbf{E}}_t\left[S_T^a\right]}\hat{\mathbf{E}}_t\left[\exp\Big((a+ik)X_T\Big)\right]\\
&= \frac{1}{\hat{\mathbf{E}}_t[S_T^a]}\hat{\mathbf{E}}_t\left[\exp\left((a+ik)\left(X_t + r(T-t) - \int_t^T \frac{\sigma_u^2}{2}du\right.\right.\right.\\
&\qquad\left.\left.\left. + \int_t^T \sigma_u d\hat{W}_u^S\right)\right)\right]\\
&= \frac{e^{(a+ik)(r(T-t)+X_t)}}{\hat{\mathbf{E}}_t[S_T^a]}\hat{\mathbf{E}}_t\left[\exp\left((a+ik)\left(-\int_t^T \frac{\sigma_u^2}{2}du\right.\right.\right.\\
&\qquad\left.\left.\left. + \int_t^T \sigma_u d\hat{W}_u^S\right)\right)\right].
\end{aligned}
$$

For convenience, we have assumed a constant interest rate r. The Cholesky decomposition with $(\hat{W}^{(1)}, \hat{W}^{(2)})$ being independent components of a two-dimensional Brownian motion under a risk-neutral measure $\hat{P}$ leads to

$$\hat{\chi}_{\ln S_T,t}(k) = c_1(k)\hat{\mathbf{E}}_t\left[\exp\left(ik\left(-\int_t^T \frac{\sigma_u^2}{2}du + \gamma\int_t^T \sigma_u d\hat{W}_u^{(1)} + \sqrt{1-\gamma^2}\int_t^T \sigma_u d\hat{W}_u^{(2)}\right)\right)\right],$$

$$\chi_{\ln S_T,t}^{(a)}(k) = c_2(k)\hat{\mathbf{E}}_t\left[\exp\left((a+ik)\left(-\int_t^T \frac{\sigma_u^2}{2}du + \gamma\int_t^T \sigma_u d\hat{W}_u^{(1)} + \sqrt{1-\gamma^2}\int_t^T \sigma_u d\hat{W}_u^{(2)}\right)\right)\right],$$

where

$$c_1(k) = \exp\left(ik(r(T-t)+X_t)\right)$$

$$c_2(k) = \frac{\exp\left((a+ik)(r(T-t)+X_t)\right)}{\hat{\mathbf{E}}_t(S_T^a)}.$$

Let $\mathcal{G}_t$ be the σ-algebra generated by the stochastic volatility process and $\bar{\mathcal{G}}_t = \sigma\left(\mathcal{F}_t \cup \mathcal{G}\right)$. Conditioning on $\bar{\mathcal{G}}_t$ and using the formula for the expectation of a log-normally distributed variable as well as the Itô-isometry yield

$$\hat{\chi}_{\ln S_T,t}(k) = c_1(k)\hat{\mathbf{E}}_t\Bigg[\exp\Bigg(ik\Big(-\int_t^T \frac{\sigma_u^2}{2}du + \gamma\int_t^T \sigma_u d\hat{W}_u^{(1)}\Big)\Bigg)\times$$

$$\hat{\mathbf{E}}\Bigg[\exp\Bigg(ik\sqrt{1-\gamma^2}\int_t^T \sigma_u d\hat{W}_u^{(2)}\Bigg)\Bigg|\bar{\mathcal{G}}_t\Bigg]\Bigg]$$

$$= c_1(k)\hat{\mathbf{E}}_t\Bigg[\exp\Bigg(ik\Big(-\int_t^T \frac{\sigma_u^2}{2}du + \gamma\int_t^T \sigma_u d\hat{W}_u^{(1)}\Big)\Bigg)\times$$

$$\exp\Bigg(\frac{1}{2}(ik)^2(1-\gamma^2)\int_t^T \sigma_u^2 du\Bigg)\Bigg]$$

$$= c_1(k)\hat{\mathbf{E}}_t\Bigg[\exp\Bigg(\frac{ik}{2}\Big(-1+ik(1-\gamma^2)\Big)\int_t^T \sigma_u^2 du\Bigg)\times$$

$$\exp\Bigg(ik\gamma\int_t^T \sigma_u d\hat{W}_u^{(1)}\Bigg)\Bigg],$$

$$\chi^{(a)}_{\ln S_T,t}(k) = c_2(k)\hat{\mathbf{E}}_t\Bigg[\exp\Bigg((a+ik)\Big(-\int_t^T \frac{\sigma_u^2}{2}du + \gamma\int_t^T \sigma_u d\hat{W}_u^{(1)}\Big)\Bigg)\times$$

$$\hat{\mathbf{E}}\Bigg[\exp\Bigg((a+ik)\sqrt{1-\gamma^2}\int_t^T \sigma_u d\hat{W}_u^{(2)}\Bigg)\Bigg|\bar{\mathcal{G}}_t\Bigg]\Bigg]$$

$$= c_2(k)\hat{\mathbf{E}}_t\Bigg[\exp\Bigg((a+ik)\Big(-\int_t^T \frac{\sigma_u^2}{2}du + \gamma\int_t^T \sigma_u d\hat{W}_u^{(1)}\Big)\Bigg)\times$$

$$\exp\Bigg(\frac{1}{2}(a+ik)^2(1-\gamma^2)\int_t^T \sigma_u^2 du\Bigg)\Bigg]$$

$$= c_2(k)\hat{\mathbf{E}}_t\Bigg[\exp\Bigg(\frac{a+ik}{2}(-1+(a+ik)(1-\gamma^2))\int_t^T \sigma_u^2 du\Bigg)\times$$

$$\exp\Bigg((a+ik)\gamma\int_t^T \sigma_u d\hat{W}_u^{(1)}\Bigg)\Bigg].$$

The derivation so far is fully general.

A.2 Ornstein–Uhlenbeck Process for Volatility

Now, the choice of the volatility process becomes important. We have chosen an Ornstein–Uhlenbeck process for the volatility σ under a risk-neutral measure $\hat{P}$:

$$d\sigma_t = \kappa^*(\theta^* - \sigma_t)dt + v d\hat{W}_t^{(1)}.$$

In order to eliminate the integral over $d\hat{W}^{(1)}$ we have to use the properties of the quadratic variation, denoted by $[\cdot,\cdot]$, see for instance Protter [36]:

$$[\sigma,\sigma]_T - [\sigma,\sigma]_t = \sigma_T^2 - \sigma_t^2 - 2\int_t^T \sigma_u d\sigma_u = v^2(T-t).$$

From this we obtain

$$\begin{aligned}
\frac{v(T-t)}{2} &= \frac{1}{2v}\left(\sigma_T^2 - \sigma_t^2 - 2\int_t^T \sigma_u d\sigma_u\right) \\
&= \frac{1}{2v}\left(\sigma_T^2 - \sigma_t^2 - 2\kappa^*\theta^*\int_t^T \sigma_u du + 2\kappa^*\int_t^T \sigma_u^2 du - 2v\int_t^T \sigma_u d\hat{W}_u^{(1)}\right),
\end{aligned}$$

which is equivalent to

$$\int_t^T \sigma_u d\hat{W}_u^{(1)} = \frac{1}{2v}\left(\sigma_T^2 - \sigma_t^2 - 2\kappa^*\theta^*\int_t^T \sigma_u du + 2\kappa^*\int_t^T \sigma_u^2 du - v^2(T-t)\right).$$

This in turn yields

$$\begin{aligned}
\hat{\chi}_{\ln S_T,t}(k) &= c_1(k)\hat{\mathbf{E}}_t\Bigg[\exp\Bigg(\frac{ik}{2}\Big(-1+ik(1-\gamma^2)\Big)\int_t^T \sigma_u^2 du + ik\frac{\gamma}{2v}\times \\
&\qquad \left(\sigma_T^2 - \sigma_t^2 - v^2(T-t) + 2\kappa^*\int_t^T(-\theta^*\sigma_u + \sigma_u^2)du\right)\Bigg)\Bigg] \\
&= \hat{c}(k)\hat{\mathbf{E}}_t\left[\exp\left(\int_t^T -\hat{s}_1(k)\sigma_u^2 - \hat{s}_2(k)\sigma_u du\right)\exp(\hat{s}_3(k)\sigma_T^2)\right],
\end{aligned}$$

where

$$\begin{aligned}
\hat{c}(k) &= c_1(k)\exp\left(-ik\frac{\gamma}{2v}\left(\sigma_t^2 + v^2(T-t)\right)\right),\\
\hat{s}_1(k) &= \frac{ik}{2}\left(1 - ik(1-\gamma^2) - 2\gamma\kappa^*/v\right),\\
\hat{s}_2(k) &= ik\frac{\gamma\kappa^*\theta^*}{v},\\
\hat{s}_3(k) &= ik\frac{\gamma}{2v},
\end{aligned}$$

and

$$\begin{aligned}
\chi^{(a)}_{\ln S_T,t}(k) &= c_2(k)\hat{\mathbf{E}}_t\Bigg[\exp\Bigg(\frac{a+ik}{2}\Big(-1+(a+ik)(1-\gamma^2)\Big)\int_t^T \sigma_u^2 du + \\
&\qquad (a+ik)\frac{\gamma}{2v}\Big(\sigma_T^2 - \sigma_t^2 - v^2(T-t) + 2\kappa^*\int_t^T(-\theta^*\sigma_u + \sigma_u^2)du\Big)\Bigg)\Bigg]\\
&= c^{(a)}(k)\hat{\mathbf{E}}_t\left[\exp\left(\int_t^T -s_1^{(a)}(k)\sigma_u^2 - s_2^{(a)}(k)\sigma_u du\right)\exp(s_3^{(a)}(k)\sigma_T^2)\right],
\end{aligned}$$

where

$$\begin{aligned}
c^{(a)}(k) &= c_2(k)\exp\left(-(a+ik)\frac{\gamma}{2v}(\sigma_t^2 + v^2(T-t))\right),\\
s_1^{(a)}(k) &= \frac{a+ik}{2}(1-(a+ik)(1-\gamma^2) - 2\gamma\kappa^*/v),\\
s_2^{(a)}(k) &= (a+ik)\frac{\gamma\kappa^*\theta^*}{v},\\
s_3^{(a)}(k) &= (a+ik)\frac{\gamma}{2v}.
\end{aligned}$$

The expectations are calculated according to Feynman–Kac using the Markov property of the stochastic volatility process (see, e.g. Karatzas and Shreve [30]), setting

$$y(k;\sigma,t,T) = \hat{\mathbf{E}}\left[\exp\left(\int_t^T(-s_1(k)\sigma_u^2 - s_2(k)\sigma_u)du\right)\exp(s_3(k)\sigma_T^2)\,\middle|\,\sigma_t = \sigma\right].$$

The solution is then obtained by replacing s_i by $\hat{s}_i$ and $s_i^{(a)}$, respectively. According to Feynman–Kac the function $y(\sigma,t,T) \equiv y(k;\sigma,t,T)$ solves the following partial differential equation for arbitrary complex coefficients s_1, s_2, s_3:

$$\frac{1}{2}v^2\frac{\partial y^2}{\partial\sigma^2} + \kappa^*(\theta^* - \sigma)\frac{\partial y}{\partial\sigma} - (s_1\sigma^2 + s_2\sigma)y + \frac{\partial y}{\partial t} = 0, \qquad \text{(A.1)}$$

with boundary condition

$$y(x, \sigma, T, T) = \exp(s_3 \sigma^2).$$

The solution of this partial differential equation is of the following form:

$$\begin{aligned} y(\sigma, t, T) &= \exp\left(\frac{1}{2}A(t,T)\sigma^2 + B(t,T)\sigma + C(t,T) + s_3\sigma^2\right) \\ &= \exp\left(\frac{1}{2}D(t,T)\sigma^2 + B(t,T)\sigma + C(t,T)\right), \end{aligned} \tag{A.2}$$

with $D(t,T) = A(t,T) + 2s_3$ and $A(T,T) = B(T,T) = C(T,T) = 0$.

Plugging this trial solution into the partial differential equation (A.1) yields

$$\frac{1}{2}v^2 y\left(D + \sigma^2 D^2 + 2\sigma BD + B^2\right) + \kappa^*(\theta^* - \sigma)y(\sigma D + B) - (s_1\sigma^2 + s_2\sigma)y$$
$$+y\left(\frac{1}{2}\sigma^2\frac{\partial D}{\partial t} + \sigma\frac{\partial B}{\partial t} + \frac{\partial C}{\partial t}\right) = 0.$$

Rearranging the terms leads to

$$\frac{\sigma^2}{2}\left(v^2 D^2 + \frac{\partial D}{\partial t} - 2\kappa^* D - 2s_1\right) + \sigma(v^2 BD - \kappa^* B + \kappa^*\theta^* D + \frac{\partial B}{\partial t} - s_2)$$
$$+\left(\frac{v^2}{2}D + \frac{v^2}{2}B^2 + \frac{\partial C}{\partial t} + \kappa^*\theta^* B\right) = 0.$$

Reading this equation as a polynomial in σ, it vanishes for all σ if and only if all three coefficients vanish. This leads to the following system of ordinary differential equations:

$$\begin{aligned} \frac{\partial D}{\partial t} &= -v^2 D^2 + 2\kappa^* D + 2s_1, \\ \frac{\partial B}{\partial t} &= (\kappa^* - v^2 D)B - \kappa^*\theta^* D + s_2, \\ \frac{\partial C}{\partial t} &= -\frac{1}{2}v^2 B^2 - \kappa^*\theta^* B - \frac{1}{2}v^2 D, \end{aligned}$$

with boundary conditions $D(T,T) = 2s_3, B(T,T) = C(T,T) = 0$.

It is possible to verify that the following expressions are solutions to this system of ordinary differential equations:

$$D(t,T) = \frac{1}{v^2}\left(\kappa^* - \gamma_1 \frac{\sinh(\gamma_1(T-t)) + \gamma_2\cosh(\gamma_1(T-t))}{\cosh(\gamma_1(T-t)) + \gamma_2\sinh(\gamma_1(T-t))}\right),$$

$$B(t,T) = \frac{1}{v^2\gamma_1}\Bigg(\kappa^* - \kappa^*\theta^*\gamma_1 - \gamma_1 \times$$
$$\frac{\kappa^*\theta^*\gamma_1 - \gamma_2\gamma_3 + \gamma_3(\sinh(\gamma_1(T-t)) + \gamma_2\cosh(\gamma_1(T-t)))}{\cosh(\gamma_1(T-t)) + \gamma_2\sinh(\gamma_1(T-t))}\Bigg),$$

$$C(t,T) = -\frac{1}{2}\ln\left[\cosh(\gamma_1(T-t)) + \gamma_2\sinh(\gamma_1(T-t))\right] + \frac{1}{2}\kappa^*(T-t) +$$
$$\frac{{\kappa^*}^2{\theta^*}^2\gamma_1^2 - \gamma_3^2}{2v^2\gamma_1^3}\left(\frac{\sinh(\gamma_1(T-t))}{\cosh(\gamma_1(T-t)) + \gamma_2\sinh(\gamma_1(T-t))} - \gamma_1(T-t)\right)$$
$$+\frac{\kappa^*\theta^*\gamma_1\gamma_3 - \gamma_2\gamma_3^2}{v^2\gamma_1^3}\left(\frac{\cosh(\gamma_1(T-t)) - 1}{\cosh(\gamma_1(T-t)) + \gamma_2\sinh(\gamma_1(T-t))}\right),$$

where $\gamma_1 = \sqrt{2v^2 s_1 + {\kappa^*}^2}$, $\gamma_2 = \frac{1}{\gamma_1}(\kappa^* - 2v^2 s_3)$, $\gamma_3 = {\kappa^*}^2\theta^* - s_2 v^2$.

Plugging these functions into (A.2) and combining them with the parameters $\hat{s}_i, s_i^{(a)}$, $(i = 1, \ldots, 3)$ and the coefficients $\hat{c}, c^{(a)}$ give the explicit form of the characteristic functions $\hat{\chi}$ and $\chi^{(a)}$.

References

1. Almgren R, Chriss N (2000/2001) Optimal Execution of Portfolio Transactions. Journal of Risk 3(2):5–39,
2. Bakshi G, Cao C, Chen Z (1997) Empirical Performance of Alternative Option Pricing Models. Journal of Finance 52(5):2003–2049
3. Bakshi G, Madan D (2000) Spanning and Derivative-Security Valuation. Journal of Financial Economics 55(2):205–238,
4. Baxter M, Rennie A (1999) Financial Calculus — An Introduction to Derivative Pricing. Cambridge University Press, Cambridge
5. Belledin M, Schlag C (1999) An Empirical Comparison of Alternative Stochastic Volatility Models. Working Paper Series: Finance and Accounting 38, Goethe-Universität, Frankfurt
6. Björk T. (1998) Arbitrage Theory in Continuous Time. Oxford University Press, Oxford
7. Black F, Scholes M (1973) The Pricing of Options and Corporate Liabilities. Journal of Political Economy 81(3):637–654
8. Borodin AN, Salminen P (1996) Handbook of Brownian Motion. Birkhäuser, Basel
9. Branger N, Esser A, Schlag C (2002) Attainability of European Path-Independent Claims in Incomplete Markets. Working Paper, Goethe-University, Frankfurt
10. Breiman L (1992) Probability. SIAM, Philadelphia
11. Cox JC, Ingersoll JE, Ross SA (1985) An Intertemporal General Equilibrium Model of Asset Prices. Econometrica 53(2):363–384,
12. Cox JC, Ingersoll JE, Ross SA (1985) A Theory of the Term Structure of Interest Rates. Econometrica 53(2):385–408
13. Cox JC, Ross SA, Rubinstein M (1979) Option Pricing: A Simplified Approach. Journal of Financial Economics 7(3):229–263

14. Cvitanić J, Ma J (1996) Hedging Options for a Large Investor and Forward-Backward SDE's. Annals of Applied Probability 6(2):370–398
15. Delbaen F, Schachermayer W (1994) A General Version of the Fundamental Theorem of Asset Pricing. Mathematische Annalen 300(3):463–520
16. Dothan MU (1990) Prices in Financial Markets. Oxford University Press, Oxford
17. Dubil R (2002) Optimal Liquidation of Large Security Holdings in Thin Markets. Working Paper, University of Connecticut, Storrs
18. Duffie D (1996) Dynamic Asset Pricing Theory. Princeton University Press, Princeton
19. Eberlein E, Jacod J (1997) On the Range of Option Prices. Finance and Stochastics 1(2):131–140
20. Esser A (2003) General Valuation Principles for Arbitrary Payoffs and Applications to Power Options under Stochastic Volatility. Forthcoming in: Financial Markets and Portfolio Management
21. Esser A, Mönch B (2003) Modeling Feedback Effects with Stochastic Liquidity. Working Paper, Goethe-University, Frankfurt, submitted for publication
22. Frey R (2000) Market Illiquidity as a Source of Model Risk in Dynamic Hedging. In: Gibson R(ed) Model Risk. RISK Publications, London
23. Frey R, Patie P (2001) Risk Management for Derivatives in Illiquid Markets: A Simulation Study. Intermediate Report RiskLab, ETH-Zentrum, Zürich
24. Frey R, Sin CA (1999) Bounds on European Option Prices under Stochastic Volatility. Mathematical Finance 9(2):97–116,
25. Frey R, Stremme A (1997) Market Volatility and Feedback Effects from Stochastic Hedging. Mathematical Finance 7(4):351–374
26. Furlan M, Pechtl A (1997) Power Options and the Weierstrass Approximation Theorem in Option Pricing Theory. Working Paper, DG Bank, Frankfurt
27. Geman H, El Karoui N, Rochet JC (1995) Changes of Numeraire, Changes of Probability Measure and Option Pricing. Journal of Applied Probability 32(2):443–458
28. Heston S (1993) A Closed-Form Solution for Options with Stochastic Volatility with Applications to Bond and Currency Options. The Review of Financial Studies 6(2):327–343
29. Kampovsky A, Trautmann S (2000) A Large Trader's Impact on Price Processes. Working Paper, Gutenberg-University, Mainz
30. Karatzas I, Shreve S (1999) Brownian Motion and Stochastic Calculus. Springer, Berlin Heidelberg New York
31. Liu H, Yong J (2002) Option Pricing With an Illiquid Underlying Asset Market. Working Paper, Washington University, St. Louis

32. Long JB (1990) The Numeraire Portfolio. Journal of Financial Economics 26(1):29–69
33. Musiela M, Rutkowski M (1998) Martingale Methods in Financial Modelling. Springer, Berlin Heidelberg New York
34. Papanicolaou G, Sircar KR (1998) General Black-Scholes Models Accounting for Increased Market Volatility from Hedging Strategies. Applied Mathematical Finance 5(2):45–82
35. Pham H, Touzi N (1996) Equilibrium State Prices in a Stochastic Volatility Model. Mathematical Finance 6(2):215–236,
36. Protter P (1995) Stochastic Integration and Differential Equations. Springer, Berlin Heidelberg New York
37. Schöbel R, Zhu J (1999) Stochastic Volatility with an Ornstein–Uhlenbeck Process: An Extension. European Finance Review 3(1):23–46
38. Schönbucher PhJ, Wilmott P (2000) The Feedback Effect of Hedging in Illiquid Markets. SIAM Journal on Applied Mathematics 61(1):232–272
39. Shephard NG (1991) From Characteristic Function to Distribution Function: A Simple Framework for the Theory. Econometric Theory 7:519–529
40. Shreve S (1997) Stochastic Calculus and Finance. Manuscript, Carnegie Mellon University, Pittsburgh
41. Tompkins RG (1999/2000) Power Options: Hedging Nonlinear Risks. The Journal of Risk 2(2):29–45
42. Zhang P (1998) Exotic Options. World Scientific Publishing, Singapore
43. Zhu J (2000) Modular Pricing of Options. Lecture Notes in Economics and Mathematical Systems 493. Springer, Berlin Heidelberg New York

Abbreviations

BS	Black–Scholes
CL	Constant liquidity
DL	Deterministic liquidity
EMM	Equivalent martingale measure
MMA	Money market account
SDE	Stochastic differential equation
SL	Stochastic liquidity
SV	Stochastic volatility
$\forall$	For all
$\exists$	There exist(s)

List of Symbols

$(\ldots)'$	Transpose of a vector
$(\ldots)^+$	Positive part of the argument in brackets
B_t	Price of the money market account at time t
b_t	Effective drift of the stock at time t under the physical measure in the SL model
C_t	Price of a European call at time t
$\bar{C}$	Maximal payoff of a capped power option
$\mathcal{C}^{1,2}, \mathcal{C}^{1,2,2}$	Class of functions of two (three) variables that are once continuously differentiable in the first argument and twice continuously differentiable in the second (second and third) argument
$c(t, \sigma)$	Diffusion coefficient of the stochastic volatility process at time t

d	Factor for the downward jump of the risky asset in the binomial tree
$\mathbf{E}_t^P$	Expectation under the measure P conditional on $\mathcal{F}_t$
F_t	Price of a contingent claim at time t
f	Arbitrary function
$\mathcal{F} = \bigcup_{0 \leq t \leq T} \mathcal{F}_t$	Filtration generated by the exogenously given sources of risk
g	Arbitrary positive function
$\mathcal{G}$	σ-algebra generated by the stochastic volatility process
$\bar{\mathcal{G}}_t = \sigma(\mathcal{F}_t \cup \mathcal{G})$	σ-algebra generated by the union of $\mathcal{F}_t$ and $\mathcal{G}$
$H, \bar{H}$	Joint density of τ and ρ_τ ($\bar{\tau}$ and $S_{\bar{\tau}-}$) under the risk-neutral measure
h	Risk-neutral density of the first hitting time τ by the process $\ln S$
$\bar{h}_t$	Risk-neutral density of ρ at time t
$I(A)$	Indicator variable of the event A
$K, \bar{K}$	Strike prices
$l(t,\sigma), l^*(t,\sigma)$	Drift of the stochastic volatility process under the physical (risk-neutral) measure

M_t, N_t	Price of the numeraires M and N at time t
$\mathcal{N}$	Distribution function of the standard normal distribution
P	Physical measure
$\hat{P}$	Risk-neutral measure
$\bar{P}, P', P''$	Measures equivalent to the physical measure
$\tilde{P}$	Equivalent martingale measure corresponding to a risky basis asset
$\breve{P}$	Equivalent martingale measure corresponding to the portfolio V
$P^{(a)}$	Measure equivalent to $\hat{P}$ with Radon–Nikodym derivative $Y = \frac{S_T^a}{\hat{\mathrm{E}} S_T^a}$
$p, \hat{p}$	Physical and risk-neutral probability for an upward jump of the risky asset in the binomial tree
Pow_t	Price of the power option at time t
Powc_t	Price of the capped power option at time t
Powd_t	Price of the powered option at time t
r_t	Deterministic short rate of interest from t to $t + dt$
$\Re$	Real part of a complex number or function

$S_t^{(1)}, \ldots, S_t^{(m)}$	Prices of basis assets at time t
$S_t^{(m+1)}, \ldots, S_t^{(n)}$	State variables at time t
$dS^{(i)} \bullet dS^{(j)}$	Instantaneous covariation between $S^{(i)}$ and $S^{(j)}$
$\bar{s}$	Lower threshold for the stock price process
u	Parameter for the upward jump of the risky asset in the binomial tree
V_t	Price of the portfolio V at time t
v	Volatility of the Ornstein–Uhlenbeck process for stochastic volatility
$W = (W^{(1)}, \ldots, W^{(n)})'$	n-dimensional Brownian motion under P
$\hat{W}, \tilde{W}, \breve{W}$	n-dimensional Brownian motion under $\hat{P}, \tilde{P}, \breve{P}$
$\mathcal{W}$	Arbitrary positive payoff adapted to $\mathcal{F}_T$
X_t	$= \ln S_t$
$x_t^{(i)}, x_t^B$	Portfolio weights of the risky basis assets and the MMA at time t
Y	Radon–Nikodym derivative on $\mathcal{F}_T$
$Z = (Z^{(1)}, \ldots, Z^{(n)})'$	n-dimensional vector of orthogonal basis martingales under P
$\hat{Z}, \tilde{Z}, \breve{Z}$	n-dimensional vector of non-orthogonal basis martingales under $\hat{P}, \tilde{P}, \breve{P}$

$\alpha = (\alpha^{(1)}, \ldots, \alpha^{(n)})'$	n-dimensional predictable process determining the measure $\bar{P}$
β	A root of the covariance matrix in the discrete setup
$\Gamma(x, z)$	Incomplete gamma function
$\Gamma(x) = \Gamma(x, 0)$	Complete gamma function
γ	Correlation between the increments of the Brownian motions
δ_{ij}	Kronecker delta, equal to one for $i = j$, equal to zero for $i \neq j$
Δ_t	Delta = numbers of the underlying asset held at time t
$\zeta_0, \bar{\zeta}_0$	Price of insurance contracts against illiquidity discounts at $t = 0$
$\eta(t, \rho), \eta^*(t, \rho)$	Drift of the stochastic liquidity process at time t under the physical (risk-neutral) measure
θ, θ^*	Long-term mean of a mean reversion process under the physical (risk-neutral) measure
κ, κ^*	Speed of mean reversion under the physical (risk-neutral) measure
$\lambda = (\lambda^{(1)}, \ldots, \lambda^{(n)})'$	Market price of risk vector

$\tilde{\lambda}, \breve{\lambda}$	n-dimensional predictable process determining the measure $\tilde{P}, \breve{P}$
μ_t	Drift of the underlying asset at time t under the physical measure
$\nu(t,\rho)$	Volatility of the stochastic liquidity process at time t
ξ_t	$= \mathbf{E}_t[Y]$
ρ_t	Liquidity process at time t
$\tilde{\rho}$	Global bound for the liquidity process
$\bar{\rho}$	Threshold for the liquidity process
Σ	A root of the covariance matrix in the continuous setup
τ	First hitting time of the threshold $\bar{s}$ by the stock price process
$\bar{\tau}$	First hitting time of the threshold $\bar{\rho}$ by the liquidity process
ϕ_t	Trading strategy of the large trader at time t in the stochastic liquidity model
$\phi^{(+)}$	Right-continuous process corresponding to ϕ
$\phi^{(c)}, \phi^{(p)}$	Contrarian (positive) feedback strategy
$\hat{\chi}(k), \chi^{(a)}(k)$	Characteristic function with respect to $\hat{P}, P^{(a)}$

List of Figures

List of Tables

Index

Lecture Notes in Economics and Mathematical Systems

For information about Vols. 1–444
please contact your bookseller or Springer-Verlag

Vol. 445: A. H. Christer, S. Osaki, L. C. Thomas (Eds.), Stochastic Modelling in Innovative Manufactoring. X, 361 pages. 1997.

Vol. 446: G. Dhaene, Encompassing. X, 160 pages. 1997.

Vol. 447: A. Artale, Rings in Auctions. X, 172 pages. 1997.

Vol. 448: G. Fandel, T. Gal (Eds.), Multiple Criteria Decision Making. XII, 678 pages. 1997.

Vol. 449: F. Fang, M. Sanglier (Eds.), Complexity and Self-Organization in Social and Economic Systems. IX, 317 pages, 1997.

Vol. 450: P. M. Pardalos, D. W. Hearn, W. W. Hager, (Eds.), Network Optimization. VIII, 485 pages, 1997.

Vol. 451: M. Salge, Rational Bubbles. Theoretical Basis, Economic Relevance, and Empirical Evidence with a Special Emphasis on the German Stock Market.IX, 265 pages. 1997.

Vol. 452: P. Gritzmann, R. Horst, E. Sachs, R. Tichatschke (Eds.), Recent Advances in Optimization. VIII, 379 pages. 1997.

Vol. 453: A. S. Tangian, J. Gruber (Eds.), Constructing Scalar-Valued Objective Functions. VIII, 298 pages. 1997.

Vol. 454: H.-M. Krolzig, Markov-Switching Vector Autoregressions. XIV, 358 pages. 1997.

Vol. 455: R. Caballero, F. Ruiz, R. E. Steuer (Eds.), Advances in Multiple Objective and Goal Programming. VIII, 391 pages. 1997.

Vol. 456: R. Conte, R. Hegselmann, P. Terna (Eds.), Simulating Social Phenomena. VIII, 536 pages. 1997.

Vol. 457: C. Hsu, Volume and the Nonlinear Dynamics of Stock Returns. VIII, 133 pages. 1998.

Vol. 458: K. Marti, P. Kall (Eds.), Stochastic Programming Methods and Technical Applications. X, 437 pages. 1998.

Vol. 459: H. K. Ryu, D. J. Slottje, Measuring Trends in U.S. Income Inequality. XI, 195 pages. 1998.

Vol. 460: B. Fleischmann, J. A. E. E. van Nunen, M. G. Speranza, P. Stähly, Advances in Distribution Logistic. XI, 535 pages. 1998.

Vol. 461: U. Schmidt, Axiomatic Utility Theory under Risk. XV, 201 pages. 1998.

Vol. 462: L. von Auer, Dynamic Preferences, Choice Mechanisms, and Welfare. XII, 226 pages. 1998.

Vol. 463: G. Abraham-Frois (Ed.), Non-Linear Dynamics and Endogenous Cycles. VI, 204 pages. 1998.

Vol. 464: A. Aulin, The Impact of Science on Economic Growth and its Cycles. IX, 204 pages. 1998.

Vol. 465: T. J. Stewart, R. C. van den Honert (Eds.), Trends in Multicriteria Decision Making. X, 448 pages. 1998.

Vol. 466: A. Sadrieh, The Alternating Double Auction Market. VII, 350 pages. 1998.

Vol. 467: H. Hennig-Schmidt, Bargaining in a Video Experiment. Determinants of Boundedly Rational Behavior. XII, 221 pages. 1999.

Vol. 468: A. Ziegler, A Game Theory Analysis of Options. XIV, 145 pages. 1999.

Vol. 469: M. P. Vogel, Environmental Kuznets Curves. XIII, 197 pages. 1999.

Vol. 470: M. Ammann, Pricing Derivative Credit Risk. XII, 228 pages. 1999.

Vol. 471: N. H. M. Wilson (Ed.), Computer-Aided Transit Scheduling. XI, 444 pages. 1999.

Vol. 472: J.-R. Tyran, Money Illusion and Strategic Complementarity as Causes of Monetary Non-Neutrality. X, 228 pages. 1999.

Vol. 473: S. Helber, Performance Analysis of Flow Lines with Non-Linear Flow of Material. IX, 280 pages. 1999.

Vol. 474: U. Schwalbe, The Core of Economies with Asymmetric Information. IX, 141 pages. 1999.

Vol. 475: L. Kaas, Dynamic Macroeconomics with Imperfect Competition. XI, 155 pages. 1999.

Vol. 476: R. Demel, Fiscal Policy, Public Debt and the Term Structure of Interest Rates. X, 279 pages. 1999.

Vol. 477: M. Théra, R. Tichatschke (Eds.), Ill-posed Variational Problems and Regularization Techniques. VIII, 274 pages. 1999.

Vol. 478: S. Hartmann, Project Scheduling under Limited Resources. XII, 221 pages. 1999.

Vol. 479: L. v. Thadden, Money, Inflation, and Capital Formation. IX, 192 pages. 1999.

Vol. 480: M. Grazia Speranza, P. Stähly (Eds.), New Trends in Distribution Logistics. X, 336 pages. 1999.

Vol. 481: V. H. Nguyen, J. J. Strodiot, P. Tossings (Eds.). Optimation. IX, 498 pages. 2000.

Vol. 482: W. B. Zhang, A Theory of International Trade. XI, 192 pages. 2000.

Vol. 483: M. Königstein, Equity, Efficiency and Evolutionary Stability in Bargaining Games with Joint Production. XII, 197 pages. 2000.

Vol. 484: D. D. Gatti, M. Gallegati, A. Kirman, Interaction and Market Structure. VI, 298 pages. 2000.

Vol. 485: A. Garnaev, Search Games and Other Applications of Game Theory. VIII, 145 pages. 2000.

Vol. 486: M. Neugart, Nonlinear Labor Market Dynamics. X, 175 pages. 2000.

Vol. 487: Y. Y. Haimes, R. E. Steuer (Eds.), Research and Practice in Multiple Criteria Decision Making. XVII, 553 pages. 2000.

Vol. 488: B. Schmolck, Ommitted Variable Tests and Dynamic Specification. X, 144 pages. 2000.

Vol. 489: T. Steger, Transitional Dynamics and Economic Growth in Developing Countries. VIII, 151 pages. 2000.

Vol. 490: S. Minner, Strategic Safety Stocks in Supply Chains. XI, 214 pages. 2000.

Vol. 491: M. Ehrgott, Multicriteria Optimization. VIII, 242 pages. 2000.

Vol. 492: T. Phan Huy, Constraint Propagation in Flexible Manufacturing. IX, 258 pages. 2000.

Vol. 493: J. Zhu, Modular Pricing of Options. X, 170 pages. 2000.

Vol. 494: D. Franzen, Design of Master Agreements for OTC Derivatives. VIII, 175 pages. 2001.

Vol. 495: I Konnov, Combined Relaxation Methods for Variational Inequalities. XI, 181 pages. 2001.

Vol. 496: P. Weiß, Unemployment in Open Economies. XII, 226 pages. 2001.

Vol. 497: J. Inkmann, Conditional Moment Estimation of Nonlinear Equation Systems. VIII, 214 pages. 2001.

Vol. 498: M. Reutter, A Macroeconomic Model of West German Unemployment. X, 125 pages. 2001.

Vol. 499: A. Casajus, Focal Points in Framed Games. XI, 131 pages. 2001.

Vol. 500: F. Nardini, Technical Progress and Economic Growth. XVII, 191 pages. 2001.

Vol. 501: M. Fleischmann, Quantitative Models for Reverse Logistics. XI, 181 pages. 2001.

Vol. 502: N. Hadjisavvas, J. E. Martínez-Legaz, J.-P. Penot (Eds.), Generalized Convexity and Generalized Monotonicity. IX, 410 pages. 2001.

Vol. 503: A. Kirman, J.-B. Zimmermann (Eds.), Economics with Heterogenous Interacting Agents. VII, 343 pages. 2001.

Vol. 504: P.-Y. Moix (Ed.),The Measurement of Market Risk. XI, 272 pages. 2001.

Vol. 505: S. Voß, J. R. Daduna (Eds.), Computer-Aided Scheduling of Public Transport. XI, 466 pages. 2001.

Vol. 506: B. P. Kellerhals, Financial Pricing Models in Continuous Time and Kalman Filtering. XIV, 247 pages. 2001.

Vol. 507: M. Koksalan, S. Zionts, Multiple Criteria Decision Making in the New Millenium. XII, 481 pages. 2001.

Vol. 508: K. Neumann, C. Schwindt, J. Zimmermann, Project Scheduling with Time Windows and Scarce Resources. XI, 335 pages. 2002.

Vol. 509: D. Hornung, Investment, R&D, and Long-Run Growth. XVI, 194 pages. 2002.

Vol. 510: A. S. Tangian, Constructing and Applying Objective Functions. XII, 582 pages. 2002.

Vol. 511: M. Külpmann, Stock Market Overreaction and Fundamental Valuation. IX, 198 pages. 2002.

Vol. 512: W.-B. Zhang, An Economic Theory of Cities.XI, 220 pages. 2002.

Vol. 513: K. Marti, Stochastic Optimization Techniques. VIII, 364 pages. 2002.

Vol. 514: S. Wang, Y. Xia, Portfolio and Asset Pricing. XII, 200 pages. 2002.

Vol. 515: G. Heisig, Planning Stability in Material Requirements Planning System. XII, 264 pages. 2002.

Vol. 516: B. Schmid, Pricing Credit Linked Financial Instruments. X, 246 pages. 2002.

Vol. 517: H. I. Meinhardt, Cooperative Decision Making in Common Pool Situations. VIII, 205 pages. 2002.

Vol. 518: S. Napel, Bilateral Bargaining. VIII, 188 pages. 2002.

Vol. 519: A. Klose, G. Speranza, L. N. Van Wassenhove (Eds.), Quantitative Approaches to Distribution Logistics and Supply Chain Management. XIII, 421 pages. 2002.

Vol. 520: B. Glaser, Efficiency versus Sustainability in Dynamic Decision Making. IX, 252 pages. 2002.

Vol. 521: R. Cowan, N. Jonard (Eds.), Heterogenous Agents, Interactions and Economic Performance. XIV, 339 pages. 2003.

Vol. 522: C. Neff, Corporate Finance, Innovation, and Strategic Competition. IX, 218 pages. 2003.

Vol. 523: W.-B. Zhang, A Theory of Interregional Dynamics. XI, 231 pages. 2003.

Vol. 524: M. Frölich, Programme Evaluation and Treatment Choise. VIII, 191 pages. 2003.

Vol. 525:S. Spinler, Capacity Reservation for Capital-Intensive Technologies. XVI, 139 pages. 2003.

Vol. 526: C. F. Daganzo, A Theory of Supply Chains. VIII, 123 pages. 2003.

Vol. 527: C. E. Metz, Information Dissemination in Currency Crises. XI, 231 pages. 2003.

Vol. 528: R. Stolletz, Performance Analysis and Optimization of Inbound Call Centers. X, 219 pages. 2003.

Vol. 529: W. Krabs, S. W. Pickl, Analysis, Controllability and Optimization of Time-Discrete Systems and Dynamical Games. XII, 187 pages. 2003.

Vol. 530: R. Wapler, Unemployment, Market Structure and Growth. XXVII, 207 pages. 2003.

Vol. 531: M. Gallegati, A. Kirman, M. Marsili (Eds.), The Complex Dynamics of Economic Interaction. XV, 402 pages, 2004.

Vol. 532: K. Marti, Y. Ermoliev, G. Pflug (Eds.), Dynamic Stochastic Optimization. VIII, 336 pages. 2004.

Vol. 533: G. Dudek, Collaborative Planning in Supply Chains. X, 234 pages. 2004.

Vol. 534: M. Runkel, Environmental and Resource Policy for Consumer Durables. X, 197 pages. 2004.

Vol. 535: X. Gandibleux, M. Sevaux, K. Sörensen, V. T'kindt (Eds.), Metaheuristics for Multiobjective Optimisation. IX, 249 pages. 2004.

Vol. 536: R. Brüggemann, Model Reduction Methods for Vector Autoregressive Processes. X, 218 pages. 2004.

Vol. 537: A. Esser, Pricing in (In)Complete Markets. XI, 122 pages, 2004.

Batch number: 10844675

Printed by Printforce, the Netherlands